T0197233

HOW WE'LL ALL BE EQUALLY RICH, FAMOUS, BRILLIANT, ETC., FOREVER

LEWIS S. MANCINI, M.D.

Order this book online at www.trafford.com
or email orders@trafford.com

Most Trafford titles are also available at major online book retailers.

Printed in Victoria, BC, Canada.

ISBN: 978-1-4269-3292-2 (sc)
ISBN: 978-1-4269-3293-9 (eb)

*Our mission is to efficiently provide the world's finest, most comprehensive book publishing
service, enabling every author to experience success. To find out how to publish your book,
your way, and have it available worldwide, visit us online at www.trafford.com*

Trafford rev. 6/22/2010

 www.trafford.com

North America & international
toll-free: 1 888 232 4444 (USA & Canada)
phone: 250 383 6864 ♦ fax: 812 355 4082

"And the last shall be first, and the first, last."
—Matthew 20:16

My Favorite Summary of This Book

This book might, sooner or later, have the following extended, full-length title: *Equal Fame, Fortune and Brainpower for Everyone, Forever – because it's all inevitable, sooner or later – Plus Everyone's Right to Painless, Physician-Assisted Suicide.*

Within the time period from the present moment to 25 years from now, de-effortizing brain stimulation (that might or might not be intensely pleasurable) will probably dramatically, or at least significantly, increase human brainpower in such ways so that (within the next 50 years or so) work skills will become so easily acquirable and diversifiable that poverty and financial stress will vanish from our world.

And within the next 25 to 50 years, via reproducible out-of-body and/or near-death experiences (OBEs and NDEs) the brain and mind will be unequivocally revealed to be two different things, with the former being clearly mortal and aging-prone and the latter being clearly immortal and ageless.

The mind will be discovered to be an indivisible mind particle (MP). Within the next 50 to 100 years, via pleasurable brain stimulation, the Internet, GPS and global mind particle circulation (GMPC), the MP of each and every past, present and future human being will become interdigitated with the MP of every other human, so that everyone will become equally well acquainted with everyone else.

And therefore everyone will be equally famous. Moreover, everyone will have equal access to the wealth (including goods and services) and knowledge (including work skills) of everyone else. So everyone from the past, present and future will be equally famous, rich, intelligent, otherwise gifted, etc., forever.

The reason why all human mind <u>particles</u>, regardless of earthly birth date (whether past, present or future) stick together for eternity is because of the forces that exist between and among us. These forces include: (1) the strong nuclear force, (2) the weak nuclear force, (3) electromagnetic attraction, (4) gravity and (5) the force of sentimental affection or <u>love</u> which all <u>conscious</u>, <u>subconscious</u> and even <u>unconscious</u> particles are (at least <u>potentially</u>) strongly motivated by. Forces 1 through 4 might actually be various manifestations of force number 5 (that is, love), which is actually a manifestation of pleasure. Hence, pleasure might be the underlying, sole constructive force within the infinitely large universe. Moreover, all particles are potential <u>mind</u> particles. And pleasure might be the singular, grand, unifying force of nature, which can reconcile all valid mathematical descriptions thereof, including possibly (?) superstring theory.

Plus, since we earthlings are doing a <u>favor</u> for the infinitely large universe by living here at all, we should each be acknowledged to have the right to self-determine <u>how much</u> of a favor we're going to do and for <u>how long</u> we're going to do it. Also explored herein is a possible interface between science and religion.

Contents

About the Author

Lewis S. Mancini, M.D., is a psychiatrist with additional background in biophysics, bioengineering and electroencephalography (EEG) technology. He has been afflicted with various learning disabilities since childhood and disabled with severe obsessive-compulsive disorder (OCD) since 1985. His two major learning disabilities are gadgetaphobia and general informational phobia. He has also been diagnosed with bipolar manic-depression since 2007.

He is a graduate of St. George's University School of Medicine (1983). And previously, in late 1974, worked briefly (as a research assistant) on "Artificial Vision for the Blind," under the auspices of Dr. William H. Dobelle in the Neuroprostheses and Artificial Organs Divisions of the Bioengineering Department at the University of Utah. Dr. Dobelle is cited in the 2005 Guinness Book of World Records under the headings of Medical Phenomena and the "earliest successful artificial eye" on page 20.

The author also won an award for "academic excellence" in EEG technology from Graphic Controls Corporation and has had six articles published in *Speculations in Science and Technology* and *Medical Hypotheses*. He also had a book titled *How Everyone Could Be Rich, Famous, Etc.* published by Trafford in 2006.

One of his goals is to play whatever role he can in the conceivable implementation of brain-stimulation-mediated learning facilitation (LF) and work skills facilitation (WF), enhancement and diversification.

A Potential Business Item

(A brief, potentially money-making idea or note
to anyone purchasing one or more copies of this book)

In case I am ever effectively treated for my severe obsessive-compulsive disorder, other emotional difficulties including bipolar manic-depression, relieved of various learning disabilities, etc., and thereby enabled to return to a lifestyle of being gainfully employed, by dint of making/having made this purchase, you might be possibly getting in on the "ground-floor level" of any publishing or medical devices company that I might be involved with in the future. Therefore, please retain proof of purchase in a secure location, for possible future reference.

Please consider that I have already had six articles published in international, peer-reviewed medical and scientific literature and have already worked in a laboratory on the research and development (R&D) of a medical device (as a research assistant on the "Artificial Vision for the Blind" project, under the auspices and guidance of Dr. William H. Dobelle).

L. S. M.

Prefatory Note 1

The possibly-apparent contradiction between (a) my inclusion in this book of the potential business item and (b) my unusually strong death wish is, actually, <u>not</u> a genuine contradiction.

If my obsessive-compulsive disorder (OCD) with its accompanying <u>horrifying</u>, <u>intrusive</u> and <u>unwanted</u> thoughts, ruminative attention-deficit disorder (ADD) and associated learning disabilities can all be effectively treated with deep brain stimulation (ref. 1) or any other treatment modility/ies, I will be happy to live here on Earth until the ripe old age of 100+ years, if that is physically possible for me to do.

Reference

1. Hall, S. S. "Brain Pacemakers." *Technology Review: MIT's Magazine of Innovation*, Sept. 2001; *104* (7): 34–43.

Prefatory Note 2: The Unconventional Use of Capital Letters throughout This Book

Throughout this book, I have used capital or upper case letters in many contexts where lower-case letters would be more conventionally acceptable. The reasons for doing this are invariably either to give special emphasis to some words or to promote egalitarian ideals. I hope this relative overuse of capital letters will cause no discomfort to the reader. Thank you.

Sincerely,

L. S. M.

UNIVERSITY AT BUFFALO
 Department of Biophysical Sciences

STATE UNIVERSITY OF NEW YORK
School of Medicine and Biomedical Sciences

18 October, 1991

To Whom It May Concern:

This is to certify that Dr. Lewis Mancini has successfully completed a program in independent study with me during the Spring semester, 1989. The objectives of this program were that of exploring various physical modalities by which a small region of the human CNS could be selectively and noninvasively stimulated.

The problem of safely and conveniently stimulating restricted regions of the brain, primarily for purposes of pain control, is one in which Dr. Mancini has long held an interest; I first spoke with him in this regard about five years ago, early in his psychiatry residency. Since then his appreciation and knowledge of the physical mechanisms likely to be involved—electromagnetic and ultrasound fields—has evolved substantially. It is my understanding that Dr. Mancini intends to pursue research in this area in conjunction with or subsequent to his completion of training in psychiatry. Alternatively, Dr. Mancini is well prepared, I believe, to undertake training in bioengineering prior to his re-entry into a psychiatric residency program in order to research and develop this specific interest in techniques of brain stimulation.

Over the course of the previous semester, Dr. Mancini reviewed the theoretical basis of magnetic fields. Clearly, a purely electromagnetic phenomenon cannot be sufficiently focused sufficiently distant from generating coils or electrodes to allow adequate selectivity in stimulating a small portion of the CNS. Electromagnetic induction, in combination with another physical process, however, seems to offer a possible method of achieving adequate focusing. One such technique is that proposed by W. Fry, based upon the use of superimposed electric and ultrasound fields, both of the same frequency. Given a finite dependence of electrical conductivity upon pressure and/ or temperature, the focused ultrasound generates partial rectification of the alternating electrical field, leading to neural stimulation in that region. The small magnitude of the conductivity's pressure/temperature coefficient makes the fields necessary to achieve stimulation (theoretically) prohibitively high, however. The basis and ramifications of Fry's proposed method were extensively discussed.

A more likely methodology growing out of the discussions held during the semester is based upon the observations of several Russian authors that sufficiently intense ultrasound alone can result in neural stimulation. Dr. Mancini proposes to couple this effect with magnetically induced currents within the brain. While the precise mechanism whereby ultrasound achieves stimulation remains to be established, it may well involve alteration of membrane properties through direct mechanical action on the neural membrane at the molecular level. It is an interesting and plausible prospect that this effect would enhance the sensitivity of the neural tissue to magnetically induced electrical currents. Were this hypothesis proved true, the ultrasound field could be used to define the region to be stimulated, with the electromagnetic field providing a mechanism for stimulation at reasonable field strength of each.

In view of these considerations, it is plausible that adequately localized deep brain stimulation could be achieved by means of an external device using ultrasound at energy levels considerably lower than those suggested by Fry. In order to pursue research along these lines, I have suggested that Dr. Mancini explore opportunities at laboratories outside the Buffalo area, since this application of ultrasonics to brain stimulation is not the primary focus of any laboratory in the city.

At the conclusion of the independent study program, Dr. Mancini prepared a paper exploring these potential techniques of brain stimulation, based upon his review of the relevant literature.

Sincerely,
(signed)
Robert A. Spangler, M.D., Ph.D.
Associate Professor, Biophysics

UNIVERSITY AT BUFFALO
Department of Biophysical Sciences
STATE UNIVERSITY OF NEW YORK
Faculty of Health Sciences

3 February, 1992

To Whom It May Concern:

Dr. Lewis Mancini is currently enrolled in Independent Study (BPH600) with me. The objective of his independent study program is that of continuing to broaden his knowledge concerning ultrasound and its possible direct effects upon CNS function, and the EEG. Building upon his previous experience and training, Dr. Mancini is eager for an opportunity to obtain practical research experience in these and related areas of research in an appropriate laboratory setting. His interest and longstanding enthusiasm concerning these investigations will be a valuable asset in the exploration of a range of potential applications of these physical modalities.

4 May, 1992

To Whom It May Concern:

This is to certify that Dr. Lewis Mancini has successfully completed the program of independent study as indicated above. During the course of the semester, Dr. Mancini continued to broaden his knowledge of the potential physical basis for noninvasive focused brain stimulation by ultrasonic energy through library research and communication with researchers in this field. His ideas concerning such direct stimulation and its possible applications are being refined and written up in a manuscript which, at present, is envisioned to form the basis of two books for which he has developed detailed outlines.

Sincerely,
(signed)
Robert A. Spangler, M.D., Ph.D.
Associate Professor of
Biophysical Sciences

Brief Book Overview or Synopsis

This book attempts to explain the following three intuitively-derived question-and-answer pairs:

1. Why human life on planet Earth is problematic but nevertheless absolutely necessary in order for the infinitely large universe to be the extremely (but *not* infinitely) wonderful place that it is.

2. How it is that you and I and every other human being who has ever been or ever will be born on Earth will soon (within the next 20 to 100 years or so) be more famous, wealthier, more "brilliant" (and otherwise more fortunate) than anyone in the history of the world has been or currently is.

3. Why painless, suffering-free, physician-assisted suicide should be readily and legally available to everyone on Earth whose circumstances meet certain criteria of a poor quality of life.

Book Summary

Life on Earth is Hell. But without this finite Hell, there could be no infinite Heaven. There are two possible routes you can take to get to Heaven:

1. Die now and probably (?) qualify to go there immediately. However, if you do not qualify (either because you haven't finished your assignment here or because you've behaved in evil ways while you've been here), you could either get sent back to (a) resume your current lifetime or (b) be reincarnated into a new and less desirable lifetime. Possibility (b) is only ever intended as both a punishment and a chance to atone for misbehavior.

OR

2. Wait around here on Earth for another 20 to 100+ years, by which time science, technology, egalitarian and altruistic philosophy will probably have fully Heavenized Hell. By that time (the time of a full return to Heavenly conditions, hence "Kingdom come"), we'll probably all be *equally* (1) famous (i.e., we'll all know each other well), (2) rich (i.e., goods and services will be so plentiful that money will be unnecessary and will be obsolete), (3) brilliantly geniustic in a well-rounded sense (i.e., learning and "working" will have been de-effortized, pleasurized and instantaneously automated) and (4) otherwise fortunate, including (5) immortal.

Heaven is not a mystical, spiritual, nonphysical place. It is a real, physical and infinitely large place where science, technology, egalitarian and altruistic philosophy are maximal and optimal. It is also a place where unwanted pain and any and all conceivable and inconceivable suffering are impossible.

Moreover, since we're all doing a colossal *favor* for the infinitely large Heavenly volume of the universe by tolerating life here in Hell, we should all have the legally acknowledged right and free (but conscientiously-considered) access to painless, suffering-free, physician-assisted suicide. Incidentally, finite Earthly Hell is the only Hell in the infinite universe. The rest of the infinite universe is nothing but (almost pure) Heaven. So, Earthly Hell is actually the arbitrarily-selected (by God?) centerpoint of the infinite universe.

The questions of whether or not God exists and if He does exist, whether or not He has the power to end all suffering, are beyond the scope of this book, but may be taken up in a sequel to this one. I'm the son of an atheistic mother and a Roman Catholic father. Since I'm a Roman Catholic myself, I find myself agreeing with my father on the question of whether or not there is a God. So, according to my belief: Yes, there is a God. However, the definition of God is beyond the scope of this manuscript, except to say that there might be more than one plausible definition of Him.

Lewis S. Mancini, M.D.

Caption for Cover Drawing

Caption for the drawing on the cover: An example of one (of many possible) schematic illustration(s) of a combined brain stimulator and mind particle circulator, detector and accelerator. CO = circulation directed outward, CI = circulation directed inward, CCC? = central circulator and consciousizer with question-answering mechanism, WLO = work or learning output modality, LWI = learning or work input modality, OS = operant stimulus mediator, OR = operant response mediator. Drawing done in 1980 by Lewis Mancini.

This book is dedicated to my beloved aunts Mary and Grace; my siblings, Nora, Douglas and William; honorary siblings, Lize, Nick and Cathie; siblings-in-law, Dick, Lynn, Debbie, Christina, Sarah, Ken and Lory; three close friends, Bill, Bob and Tom; my two grade-school role models, Don and Karen; and the six individuals who have given me my six "biggest breaks" (so far, anyway), Herbert C. B., C. Timothy G., William M. H., David F. H., Robert A. S. and Thomas J. W.

It is further dedicated to my late parents, Shirley and Tony; grandparents, Alice, Mary, Louis and Eugene; honorary grandparents, Charlotte and Ellsworth; and two favorite teachers, Miriam Goldeen and Louise Amelia Braun-Jameyson (to the latter of whom there is a special dedication included herein).

Early photo of Mary:
B.S. in Education, Buffalo State College;
M.S. in Education, State University of NY at Buffalo; Certified Teacher of the Deaf,
Canisius College and State University of NY at Buffalo

Early photo of Grace:
B.S. in Education, Buffalo State College;
Certified Teacher of the Deaf,
Canisius College and State University of NY at Buffalo

Recent photo of Robert:
M.D., Harvard; Ph.D. in Biophysics, State University of NY at Buffalo

The author (center, front) and his family of origin, circa 1957 (age 6), encircled by (brother) Douglas, (mother) Shirley, (sister) Nora, (father) Anthony and (brother) William.

1968 photo of Louise Jameyson, ceramics teacher and head of the Art Department at the Park School of Buffalo, in Snyder, New York, between 1956 and 1978.

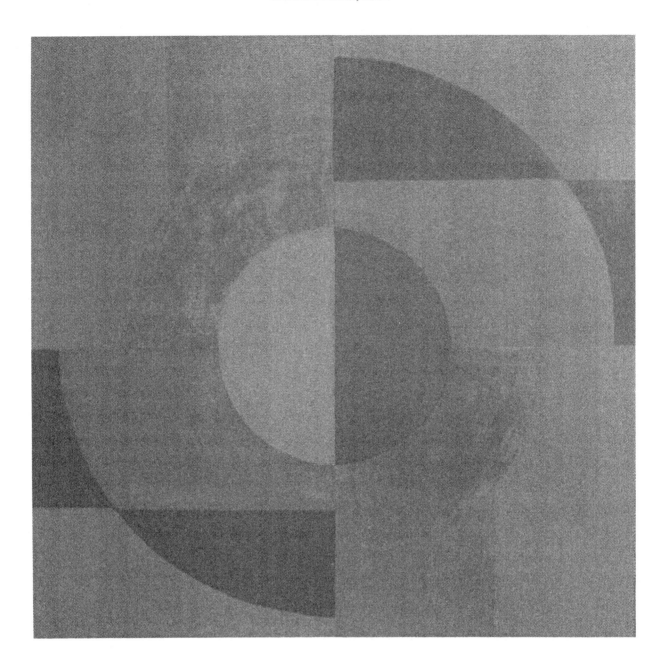

"Secret Center"
by
Louise Amelia Braun-Jameyson
1986

Acrylic on Masonite, 28" x 28"

Special Dedication to
Louise Amelia Braun-Jameyson
and Her Painting Titled
"Secret Center"

Mrs. Louise Jameyson's lifetime here on planet Earth spanned from 1907 to 1997. But since I don't believe in death as an end point of consciousness, it makes more sense to refer to her in the present tense, except when reference is being made to events which occurred in the past.

I met her in 1958 in a context of her being head of the art department and the ceramics teacher at the grade school I attended between the ages of 7 and 17. She was friendly and down to earth, yet outspokenly matter of fact, direct, assertive and bossy in a motherly, "Look, I'm old enough to be your grandmother" way. She was also simultaneously creative and *practical* about how to harness and take to fruition her own as well as any and all of her students' artistic impulses.

She was permissive enough to allow low-volume conversation among us students as we worked on our respective clay, metal or wooden objects of art. But she would raise her voice angrily and order us all to quiet down any time she thought the conversational activity had replaced the potential artwork as the central focus of attention.

I felt so comfortable with her (as though she were one of my elderly aunts) that one day, to my own surprise, when I was tired of hearing her tell us to quiet down, I said, "You know what, Jamey?" (Her nickname is Jamey) She said, "No, what?" I said, "Right now you're just being a big jerk!" (She was always a rather large person.)

Guess what happened? She got up from her bench-seat, grabbed my left ear and led me to the door of the art studio, whereupon she ejected me with the following stern advice: "Come back when you're ready to apologize." I was so embarrassed that I told no one (who hadn't witnessed it with their own eyes and ears) about the incident.

I considered complaining to my parents, but doing so would have only led to conflict and unhappy results for everyone, especially me and my family. So, realizing that she was basically like a second mother and that my verbal behavior had been out of line, I apologized the next morning. And Jamey and I have been friends ever since.

So, to me, she is a specially good person. It seems I was the last person she spoke to before she passed on. Her last words to me were simply "Well, bye-bye, dear."

Although she was deeply religious in a "born Lutheran but converted to Roman Catholic sense," she never expressed any particular scientific (such as astrophysical) ideas or hypotheses that might have related to the interface between science and religion.

Nevertheless, the uncanny and highly coincidental title of Jamey's 1986 painting, "Secret Center," suggests that she was feeling and perhaps even thinking subconsciously along the same lines as I was feeling and thinking. "Secret Center" is a study in shades of red and blue that resembles or suggests a Christian cross superimposed over a moving planet-like orb that could be taken to symbolize planet Earth along with *the Lord's cosmically-center-stage sacrifice* superimposed over it. Hence, it might be interpreted as both a religious and a scientifically suggestive work of art.

In any case, since the painting and its title could symbolize the human condition of *possibly* being center-stage to the infinite cosmos, despite our status as such being *actively* and incessantly hidden and kept a *secret* from us by an infinite horde of enviously angry extraterrestrial souls, it seems fitting that this painting should be acknowledged to have a possibly appropriate extended, alternate title of "Secret Center: How Famous and Glorious We Earthlings Will All Accurately Feel and How Valuable to the Universe We'll Realize Our Suffering Has Been When We're All Finally Allowed to Have the Truth about Our Center-Stage Status in the Infinite Cosmos."

Consequently, it is also appropriate to give Jamey, the creator of the painting, this special dedication. I hope you like it, dear Jamey.

First Prologue: Physician-Assisted Suicide

There are three reasons why painless, suffering-free suicide should be available to all of us earthlings on conscientiously-considered demand.

(1) Mental and/or physical pain and suffering are essentially evil and should be eliminated altogether if they are too intense to be minimized to tolerably comfortable levels.

(2) If we can realize that we are being kind and merciful to our beloved pet dogs, cats, horses, etc., by using the option of euthanasia when *their* quality of life drops dramatically, then we should be able to do the same for us *humans* when and if we feel that the quality of *our own* respective life has dropped to an unacceptably low level.

(3) As explained in this book, we are (a) making a sacrifice, (b) performing a service and (c) doing a favor for the infinite universe at-large by living our earthly lives at all. Therefore, we should be acknowledged to have the right to self-determine *when* we feel we have done as much of (a) through (c) as we are *willing and going to do*.

And regardless of whether you believe Jesus Christ is God (as I do) or you simply believe that He was an extremely inspired and inspiring historical figure, the *three years' time* that He took to proactively accomplish His earthly (and, I believe, Heavenly) teaching purpose (ref. 1) should be time enough for any human being, no matter how inspired they are, to accomplish *their* purpose or assignment here on Earth.

So, if any person, *regardless* of whether or not they are deemed to have any mental and/or physical illness *consistently* (for three years' time) expresses a wish to terminate their life, that person should be helped to do so, painlessly and without suffering, with the assistance of a physician.

If a person is apparently or claims to be so mentally and/or physically ill (for example, in terrible pain) that they cannot bear to wait the full three years, then, if three physicians agree that they should be allowed and helped to terminate their own life immediately, then the suffering person should and *would* be effectively assisted (by the three physicians) in doing so.

The significance of the number three (in terms of physicians required) is that, according to my own psychiatric knowledge and, more importantly, firsthand *experience* (from my February 2008 nervous breakdown), the law required only three physicians' agreement and certification that a person needs to be hospitalized *against their own will* in order for that involuntary hospitalization to occur.

Furthermore, if any person who feels their mental and/or physical pain and suffering are intense enough to humanely preclude waiting the full three years' time, but they *cannot find* three physicians to support and implement euthanasia immediately as the person desires, then at least, they should be kept comfortable mentally and physically (as much as possible) via the *full gamut of all* known pharmacologic agents (including painkillers and euphoriants) and other treatment modalities (such as brain stimulation) until the three years have elapsed and the person's *necessarily consistently-expressed* death wish can be fulfilled in accordance with the three-year waiting period criterion and stipulation. Perhaps needless to say, the age of any person expressing a death wish should always be considered completely irrelevant to the question of when the wish would be facilitated and fulfilled.

Reference

1. Crystal, D. *The Cambridge Encyclopedia*: Cambridge University Press, second edition, 1994: 590.

Second Prologue: Everyone Being Equally Famous

The hypothesis that each person is simply a conscious particle trapped inside a relatively large, problematic human body, is put forward herein. If, through reproducible out-of-body experiences (refs. 1–3), you could extricate each person-particle from our respective body-prison, then the world's entire human population (numbering seven billion or so) could literally be held in the palm of any human-sized hand.

If you were holding seven billion or so people-particles in your hand, it is easy to comprehend how *each* of the particles could interact with and become well acquainted with all (each and every one) of the other person-particles. This could be accomplished via particle detectors and accelerators such as are *currently* in use in contexts of experiments on the subject of theoretical particle physics. So, the idea of every person on Earth being well acquainted with every other one is not nearly as farfetched as it might seem at first glance.

In fact, my prediction is that it will happen, sooner or later, probably within the next 50 years. And this global person-particle inter-circulation and mutual interaction will probably involve not only all of the person-particles taken out of currently alive bodies, but rather all such particles taken out of all human bodies that have ever and will ever be born on Earth. This is because all people-particles are probably immortal and all of those who have already "lived" and "died" probably reside together with all of those which have yet to be born, "live" and "die" (from *their* bodies).

So, all past, present and future earthly person-particles will probably eventually interact, on an instantaneously inter-circulating basis, with all of the rest of the earthly person-particles. Hence, everyone will be approximately equally well acquainted with everyone else and, therefore, *equally famous*. A fuller discussion of this topic is presented below.

References

1. Kotler, S. "Extreme States." *Discover*, July 2005; *26* (7): 60–67.
2. Hoppe, C. "Controlling epilepsy." *Scientific American Mind*, June/July 2006; *17* (3): 62–67.
3. Bosveld, J. "Soul search: Can science ever decipher the secrets of the human soul?" *Discover*, June 2007; special issue: 46–50.

Third Prologue: How Poverty, Financial Stress, Etc., Might Be Alleviated:

How Poverty, Financial Stress, Pain, Anxiety, Depression, Insomnia, Cancer, Cardiovascular Disease, Blindness, Deafness, Paralysis, Obesity, Nausea, Choking, Etc., Might Be Alleviated (or Effectively Treated) by Brain Stimulation, Electromagnetism, Sound (Waves) (e.g., Ultrasound), Particle Beams, Etc.

by
Lewis S. Mancini

June 1, 2005

Summary or Abstract

Abstract – Poverty and financial stress might be alleviated by **(preferably) noninvasive electromagnetism (EM or E)**, **sound (S)**, for example, **ultrasound (US)**, **Particle-Beam (PB or P)** mediated, **possibly perceptibly pleasurable brain stimulation (PBS)** (hence, overall: **EUSPBS** or **ESPBS**) that might enable virtually anyone to **(learning-facilitatedly, LF)** learn new job skills quickly and seemingly almost effortlessly **(working-facilitatedly, WF)**, so that (preferably) high-paying scientific, high-technology and other kind(s) of employment would be within reach for (almost-?) everyone. Headaches (and all other kinds of aches), (back and all other kinds of) pain, anxiety, depression and nausea might also be relieved by ESPBS. Insomnia, blindness and deafness might be alleviated by stimulation, respectively, of the brain's sleep, vision and hearing subserving pathways. Paralysis might be alleviated by noninvasive, wireless, electromagnetically-/acoustically-/particle-beam-mediated conveyance/transmission of the brain's motor-cortex-initiated impulses in order to move/change body position via stimulation of the skeletal muscles. Cancer might be effectively treated by noninvasive electromagnetic/acoustical (e.g., ultrasonic) or particle-beam-mediated, highly focused ablation. Cardiovascular disease might be treated via noninvasive dissolution of blood clots, thromboembolic phenomena and atherosclerotic plaques, cauterization of leaky, hemorrhagic areas and noninvasive blood-pumping. Obesity might be alleviated via stimulation of the brain's food-satiety center. And choking might be relieved by relaxing (antispasmodic), inhibitory stimulation of the (bilateral) larynx-controlling areas of the brain's motor cortex.

How Poverty and Financial Stress Might Be Alleviated

It might be helpful to bear in mind, in the context of this section's perusal that if poverty and financial stress could be markedly diminished, then the broader category of stress might also be minimized. And if the broader category of stress could be minimized, then the burden of stress-related mental and physical/bodily illness and infirmity might also (i.e., consequently) be minimized.

Now, turning attention to the potential practicalities of poverty and financial stress alleviation, let us proceed as follows. Whenever a person engages in learning or working (as opposed to idle rumination), this individual's brain, neural and muscle activity or activation patterns change in detectably characteristic and monitorable ways (1–20). Let us call these changes "learning-" and "working-linked characteristics," **LLCs** and **WLCs.** What is being proposed here is that LLCs and WLCs be used as the **necessary** and **sufficient TURN-ON** and **STAY-ON signals** for a (preferably noninvasive) pleasure-mediating brain stimulator (or, synonymously for this context, brain or neural pacemaker, neurostimulator, neuromodulator, neuroactivator or neuroprosthesis).

Accordingly, **the student/worker would receive the pleasure (pleasurable brain stimulation) if and only if, whenever and only whenever and for as long as and only for as long as this individual were (engaging in a process of) learning and/or working, as necessarily indicated and (brain-stimulation-circuit-drivingly) signaled by this person's emission of one or more LLCs and/or WLCs.** The subject matter being learned/studied and/or work skills being learned/exercised (virtually regardless of its/their nature) might become intensely interesting to the student/worker because of its/their being simultaneous (or alternated) with pleasurable brain stimulation. Hence, anyone might readily and quickly develop new interests and areas of knowledge/expertise at any time (21).

It would be necessary to focus on those brain sites or pathways that subserve feelings of well-being and self-confidence (i.e., pleasures of the mind or mental pleasures) (22) as opposed to those subserving physical or bodily pleasures, such as sexual or gustatory arousal or gratification. This would be because physical pleasures tend to distract attention away from rather than potentiate learning and/or working-related processes.

It might be a good idea to use both **electromagnetic (EM)** (fields, waves or lasers) and **sound/sonic/acoustic (ultrasonic, US,** or, possibly, **infrasonic, shock, audible** or otherwise describable kind(s) of acoustical waves, fields [23–25] or **sasers**— please see sasers' description below [26]) phenomena. The focused sound waves (probably ultrasound) would possibly serve to neuroanatomically define the target region in the brain by mechanically stretching, deforming or twisting and thereby opening up cell-membrane channels in the neural area(s) being focused on. And the focused electromagnetism would possibly provide the (probably ionic) **current-driving mechanism** (27).

"Sasers" can be described as follows. A saser (acoustically analogous to a laser) might be characterized as "bright sound" or "a laser that's made from sound," expressed here in acronym form that abbreviatedly represents the analogous-to-laser concept of "sound amplification by stimulated emission of sonic/acoustic (rather than electromagnetic) radiation." Sasers are currently in early experimental and developmental stages. A diverse assortment of designs are in the process of being conceived and implemented in preliminary ways (26). Both lasers and sasers, because of their potential to facilitate **convergent beam therapy**, might be expected to be helpful with the focusing aspect of targeted pleasurable brain stimulation.

LLCs and WLCs might be spectral-analyzed electroencephalographic (EEG), functional magnetic resonance imaging (fMRI), magnetoencephalographic (MEG), evoked-response potentials (EP), electromyographic (EMG), galvanic skin response (GSR), near-infrared brain-scanning (NIBS), thermal imaging (TI), or some other kinds of phenomena that identify the occurrence of learning and/or working (1–20, 28, 29).

Transcranial magnetic stimulation (TMS) (22, 30) might (?) emerge as the method of choice for getting focusable electromagnetism into the brain's mental-pleasure- (sense-of-well-being-and-self-confidence-subserving) centers and pathways. And the obstacle of the skull bone (a relative obstacle to sound but not at all to electromagnetism) might be bypassable by means of so-called "time-reversal mirrors" (TRMs) that can be described as follows (31–33).

Acoustic time-reversal mirrors are explained essentially as follows: "… a source emits sound waves. … Each transducer in a mirror array detects the sound arriving at its location and feeds the signal to a computer … each transducer plays back its sound signal in reverse in synchrony with the other transducers. The original wave is recreated, but traveling backward, retracing its passage back through the medium … refocusing on the original source point."

A relevant observation, as expressed by Mathias Fink (31, p. 97) is that "porous bone in the skull presents an energy-sapping challenge to focusing ultrasound waves on" a relatively small neuroanatomic target (such as a pleasure subserving center, site or pathway). However, "a time-reversal mirror with a modified playback algorithm can nonetheless focus ultrasound (right) through skull bone onto a small target." Time-reversal mirrors could possibly be used to focus ultrasound, electromagnetism, electron-, ionic-, such as proton-, or other, such as atomic or molecular kind(s) of particle beams onto any small (or large) target(s) in the brain (or anywhere else in the body).

Consequently, virtually any person, by means of pleasurable brain stimulation, could possibly very quickly become knowledgeable and competent in relation to virtually any (preferably high-paying) work skills. And **poverty and financial stress might** consequently **be expected to become problems of the past if virtually anyone could readily be enabled to do virtually any (kind of) job at almost any time.** At the present time, many high-technology job vacancies "go begging" and do not get filled. That problem might be solved as explained above.

What is being suggested here is really just an adaptation of the late psychologist B. F. Skinner's **operant conditioning**. The only real differences are as follows: Instead of the operant response being a behavioral one (such as pressing a lever or pushing a button), it would be a neurophysiological one (i.e., the emission of LLCs and/or WLCs), coupled with the learning and working experiences (and behaviors) which these LLCs/WLCs signify the occurrence of. And instead of the operant stimulus being an externally tangible one such as food, drink or a means of escape from confinement, it would be a purely internally or intrinsically tangible one: pleasurable brain stimulation, which would be a direct mental reward or reinforcer.

How Headaches, Pain, Depression and Nausea Might Be Effectively Relieved

The phenomenon of reciprocal inhibition consists in the mechanism underlying the observation that pleasurable brain stimulation inhibits all kinds of unpleasant mental and physical states (depression, anxiety, anger, pain, etc.) (34–37). Conversely, unpleasant components of a person's subjective experience can observably disrupt what would otherwise be pleasant mental and physical states (contentment, self-confidence, bodily comfort, etc.). So, reciprocal inhibition consists in the notion that pleasure inhibits pain and vice versa.

The reciprocal inhibitory relationship between pleasant and unpleasant (mental and/or physical) states has its basis in (a) the interplay between neuroanatomic structures and pathways and (b) neurophysiological processes within the brain and the central nervous system (CNS) as a whole. By means of this phenomenon (of reciprocal inhibition), aches, pains, depression, anxiety, and nausea could all be inhibited, hence, alleviated by pleasurable brain stimulation.

Alternatively, by using different stimulation parameters and by targeting the unpleasant-state-subserving site(s)/pathway(s) (themselves) in the brain with inhibitory stimulation, the unpleasant states might be equally well/markedly alleviated, hence treated.

How Insomnia, Blindness and Deafness Might Be Alleviated

Insomnia, blindness and deafness could possibly be alleviated/minimized by focused noninvasive brain/neural stimulation of the respective sleep-/vision-/hearing-mediating sites and pathways (such as might be neuroanatomically determined or ascertained via fMRI or some other function-determining modality) such as might be targeted within the brain/nervous system, as suggested by the work of William Dobelle (who is cited in the 2005 Guinness Book of World Records for implementation of the earliest successful "artificial eye" for the blind and under whose auspices I worked briefly as a research assistant in late 1974) and others (38–40).

We were working with flat Teflon arrays of platinum-alloyed electrodes that were placed over the primary visual cortex (at the back surface of the brain), underneath a flap of skull bone. **Noninvasive stimulation might be expected to be superior to** (that kind of) **a surgically invasive approach** for at least two reasons: with a noninvasive approach one would have:

1. much less reason to worry about the possibility of wires (that are passing through the scalp and skull bone) potentially serving as pathways for possible infectious invasion of the meninges and the brain itself and
2. there would be much more (i.e., virtually unlimited) mobility of the stimulation target areas as compared with the situation of positionally-fixed surgically-implanted electrode arrays. And even if you use **wireless**, implanted electrodes, you still have this problem of (at least) relative immobility of targeted stimulation sites.

Moreover, with (highly movable) noninvasive particle-beam/electromagnetic/ultrasonic (or other acoustic) stimulation, it would be possible to activate any site(s) and pathway(s) in the sleep-, vision-, hearing-subserving and any and all other neuroanatomic systems, as opposed to being limited (positionally speaking) to surgical electrode-implantation sites.

How Cancer and Cardiovascular Disease Might Be Effectively Treated

Returning to the insight of Mathias Fink, here might be another pertinent instance of quoting him: "Time-reversal mirror(s)" (coupled) "with a modified playback algorithm could (noninvasively) focus USs" (or EM, electron-, ionic such as proton- or other kinds of particle beams) right "through" (the) "skull bones" (and) "onto" any (benign or malignant) "tumor" or cancer cells located (anywhere) in the brain or anywhere else in the human body (31, p. 97). In support of this contention, let us consider that radiofrequency ablation is already being used to destroy cancer cells inside of living patients.

So, focused US, other kinds of acoustic phenomena such as infrasonic, audible or shock waves' EM (perhaps mediated by sasers or lasers) or particle beams could possibly be used to kill cancer (and/or benign but problematic neoplastic) cells whenever and wherever they exist (41). Furthermore, speaking from a therapeutic viewpoint of anti-neo-angiogenesis or anti-neo-angiogenically-speaking, these modalities might also be used to therapeutically-destructively target the blood supplies of cancer (or benign neoplastic but problematic) cells or cell clusters (42).

Focused EM and/or acoustic phenomena (e.g., US or shock waves) perhaps in the form of lasers, sasers and/or particle beams might also be used to dissolve and clear away blood clots/thromboembolic phenomena and atherosclerotic plaques (43) as might be blood-flow-obstructingly located anywhere in the cardiovascular system. These same modalities, if mediated via different physical stimulus parameters, might also be useful for cauterizing or sealing off leaky or hemorrhagic areas anywhere within the circulatory system. Moreover, these modalities might be used to effectively pump and circulate blood throughout the entire body, thereby affecting the role of a surgically-noninvasive "artificial heart."

How Paralysis Might Be Relieved or Effectively Treated

Paralysis might be alleviated by wireless particle-beam-electromagnetically- or acoustically-mediated transmission of cerebral-motor-cortex-initiated-and-sustained-movement subserving impulses such as might be conveyed in order to initiate and sustain task-oriented patterns of contraction and relaxation of the task-appropriate skeletal muscles. There has already been considerable success or progress in the implementation of communication of monkeys' and humans' movement-intentions to computer cursors, robotic arms, various other external devices and even their own (externally assisted) paralyzed limbs (44, 45).

How Obesity and Anorexia Could Be Minimized

Obesity could be minimized by a) stimulating the satiety- (as opposed to the appetitive or appetite-) subserving sites/pathways of the brain (including, possibly, the ventromedial hypothalamic nucleus in the tuberal region - ?) (46) or by b) inhibitingly (inhibiting by virtue of the choice of stimulus parameters used in) deactivating the appetite-subserving sites/pathways in the brain (including, possibly, "portions of") the lateral hypothalamic nucleus - ?) (46). Conversely, anorexia could possibly be treated in neuroanatomically opposite, but functionally analogous, brain-stimulative (excitatory or inhibitory, depending on stimulus parameters used) ways.

How Choking Could Be Readily and Reliably Alleviated

Choking could possibly be alleviated by a) inhibitingly stimulating the clenched larynx by applying an electromagnetic, acoustic device directly over the laryngeal area of the neck, b) abruptly and sharply stunning the laryngismus-afflicted larynx into a kind of "hands-off/caught-off-guard" or startled, paradoxic, neuromuscularly relaxed state, by means of a noninvasive application (directly over the laryngeal area) of a device such as a more-powerful-than-usual transcranial magnetic stimulator (TMS) device (47) or by c) noninvasively, inhibitingly stimulating the laryngeal motor cortical areas of the brain (directly over the tops of both ears) (48) while using appropriate stimulus parameters in such ways as to momentarily relax the larynx and then either 1) instantaneously induce either projectile vomiting of the food bolus or other object(s) obstructing the trachea or 2) instantaneously allow descent of the obstacle into the lungs, from which it could subsequently be suctioned out, after the choking emergency situation were alleviated.

Vagus or vagal nerve stimulation (VNS) together with pleasurable brain stimulation might also prove helpful in relieving a choking emergency, because VNS's parasympathetic energy-conserving (relaxing) properties (49) (together with the consciousness-maintaining effects of pleasurable brain stimulation) might tend to relax and thereby relieve the spasmodic condition (i.e., laryngismus) that is associated with choking.

Incidentally, relevant to the overall subject of (possibly-perceptibly-) pleasurable brain-stimulation-mediated learning facilitation, VNS has been found to have some possibly significantly memory-enhancing effects (50, 51).

Conclusion

In the context of the potential economic scenario depicted above, with virtually everyone being/becoming much more productive and work-skills diversifiable than we currently are (via LLC/WLC-contingent pleasurable brain stimulation, that is) via a scenario that might be described in terms of B.S. or ESPBS-mediated learning facilitation (LF) and work skills facilitation (WF), not only might we (all of us) humans be unlikely to suffer from either financial stress or poverty but also we might become the beneficiaries of markedly increased (a) collective-global as well as (b) per-capita wealth.

Consequently, **the cost of frequent** (weekly or monthly?) **whole-body scanning** (52) for cancers, atherosclerotic plaques and other (manifestations of) diseases **could be** effectively (relative to [a] and [b]) **driven downward** so low as to facilitate its being/becoming readily accessible to and affordable by the entire population. Such increased accessibility and affordability might bode tremendously well for both public as well as individual-personal health (53).

The essence of what might make the differences between/among the potential(s) of EM, sound/ultrasound (SUS) (54) and particle beams (PBs) (31, p. 97) to a) **detect and image** cancers, atherosclerotic plaques, etc., b) **stimulate or activate** brain/neural, muscular and other kinds of tissues and **c) extirpate or ablate** these disease-mediating entities might effectively consist in **differences in stimulus** (i.e., input) **parameters** of these modalities.

In summary, poverty, financial stress, pain, anxiety, depression, etc. (as listed in the title), might all be effectively alleviated/minimized by the technological methods of approach suggested herein or by some other technologies (that might or might not be noninvasive) that are yet to emerge as being appropriate and useful for the applications suggested in this context. Nanoscale robots or nanobots (55–58) and genetically-engineered therapeutic viruses or viral entities (45) that could travel throughout the bloodstream (or, conceivably, even outside of it) might (?) be examples of health-promoting, relatively nonhazardous modalities that might potentially controvert the notion that noninvasive treatment modalities are preferable to invasive ones. **In any case, with there being an apparent superabundance of new technologies emerging at seemingly always accelerating rates, one cannot safely or securely place all of one's confidence in any one technology (and new applications of already invented or discovered technologies) (59).**

Acknowledgments

The author wishes to acknowledge with gratitude the encouragement and support of Herb, Mary, Grace, Robert, C. Timothy, Gloria, Michael, Cathie, Nick, Joani-Erica, Anna, Tom, Tom, Robert J., Patricia, Mike, David, William, Akhlesh, Betty, Amanda, Denise, William, Shirley, Tony, Mary, Louis, Alice, Eugene, Charlotte, Ellsworth, Ray and Bernice.

References

1. Gevins A. S., Morgan N. H., Bressler S. L. et al., Human neuroelectric patterns predict performance accuracy. *Science* 1987; 235: 580–585.
2. Williamson S. J. How quickly we forget – magnetic fields reveal a hierarchy of memory lifetimes in the human brain. *Science Spectra* 1999; 15: 68–73.
3. Walgate J., Wagner A., Buckner R. L., Schacter D., Sharpe K., Floyd C. Memories are made of this. *Science & Spirit* 1999; 10, 1:7.
4. Connor S. Thanks for the memory. *The World in 1999*. The Economist Publications, 1999: 110–111.
5. Goetinck S. Different brain areas linked to memorization. *The Buffalo News*, final edn., Sun., June 13, 1999, Science Notes: H-6.
6. Hall S. S. Journey to the center of my mind, brain scans can locate the home of memory and the land of language. They may eventually help to map consciousness. *The New York Times Magazine*, June 6, 1999; section 6: 122–125.
7. Neergaard L. Studies shed new light on memory/studies take close look at brain's memory process. *The Buffalo News* 1999; Fri., Aug. 21: A-10.
8. Sullivan M. M. (Editor). Task-juggling region in brain pinpointed. *The Buffalo News* 1999; Sun., May 23: H-6.
9. Fox M. Test on rats turns thought into action. *The Buffalo News* 1999; Sun., June 27: H-6.
10. McCrone J. States of mind, learning a task takes far more brainpower than repeating it once it's become a habit, could the difference show us where consciousness lies, asks J. M. *NewScientist*, March 20, 1999; 161: 30–33.
11. Gaidos S. Written all over your face, even the coolest criminal can't hide a guilty countenance. *NewScientist*, March 12–18, 2005; 185, 2490: 38–41.
12. Carmichael M. Medicine's next level – Improving the memory. *Newsweek*, Dec. 6, 2004; CXLIV, 23: 44–50.

13. Buderi R. The deceit detector. You didn't lie — your prefrontal cortex did. And Britton Chance is developing infrared-based brain imaging to catch it in action. *Technology Review, MIT's Magazine of Innovation*, June 2003; 106, 5: 66–69.

14. Osborne L. Savant for a day. Allan Snyder claims he can turn on a person's inner Rain Man, and then turn it off again, with the flick of a switch. All it takes is a strange set of electrodes and a radical new theory of autism, genius and the human brain. *The New York Times Magazine*, June 22, 2003; section 6: 38–41.

15. Thompson C. The lie detector that scans your brain. *The New York Times Magazine*, Dec. 9, 2001; section 6: 82.

16. Huang G. T. Mind-machine merger. A $24 million government initiative is jump-starting researchers' efforts to link brains and computers. The new push could yield thought-controlled robots, enhanced perception and communication – and might even make you smarter. *Technology Review, MIT's Magazine of Innovation*, May 2003; 106: 38–45.

17. Damasio A., Damasio H., Andersen R., Gazzaniga M. S., Haseltine E., Koch C., Kuiken T., LeDoux J., Tallal P., Gould E., Skayles J., Rose S., Hauser M. D., Sejnowski T. Neuroscience, Top scientists pinpoint the critical developments of the last 25 years and predict wonders yet to come. *Discover*, May 2005; 26, 5: 72–75.

18. Tancredi L. R. The new lie detectors. *Scientific American*, May 23, 2005; 16, 1: 46–47.

19. Wilson J. Why we laugh. *Popular Mechanics*, March 2003: 40–41.

20. Fox D. S. The inner savant. *Discover*, Feb. 2002; 23, 2: 44–49.

21. Mancini L. Brain stimulation to treat mental illness and enhance human learning, creativity, performance, altruism and defenses against suffering. *Medical Hypotheses 1986*; 21: 209–219.

22. Breuer H. A great attraction: magnetically stimulating the brain could lift depression and perhaps even boost creativity, but questions remain. *Scientific American* (special on the Mind), July 25, 2005; 16, 2: 54–59.

23. Fry W. J. Electrical stimulation of brain localized without probes – theoretical analysis of a proposed method. J. Acoust. *Soc. Am.* 1968; 44: 919–931.

24. Fry F. J. (brother of the late W. J. Fry; please note the preceding reference), of the Indianapolis Center for Advanced Research, Indianapolis, Indiana, USA, personal communication, May 27, 1987.

25. Spangler R. A., Assoc. Prof., State Univ. of NY at Buffalo, Depts. of Physiology and Biophysics: personal communication, Oct. 18, 1991.

26. Watson A. Pump up the volume, what lasers do for light, sasers promise to do for sound – once you can work out the best way to build one. *NewScientist*, March 27, 1999; 161: 36–40.

27. Spangler R. A. of the Dept. of Biophysical Science and Physiology, School of Medicine and Biomedical Sciences, State Univ. of New York at Buffalo, Buffalo, New York, USA, personal communication, late 1999.

28. Legrand L. N., Iacono W. G., McGue M. Predicting addiction: behavioral genetics uses twins and time to decipher the origins of addiction and learn who is most vulnerable. *Scientific American*, March–April 2005; 93, 2: 140–147.

29. Niedermeyer E., Lopes da Silva. Electroencephalography, Basic Principles, Clinical Applications and Related Fields: Urban & Schwarzenberg, 1987: 232.

30. George M. S., Belmaker R. H. Transcranial Magnetic Stimulation in Neuropsychiatry: American Psychiatric Press, 2000: 298 pages.

31. Fink M. Time-reversed acoustics. *Scientific American* 1999; 281: 91–97.

32. Fink M., Prada C. Ultrasonic focusing with time-reversal mirrors. Advances in Acoustic Microscopy Series. Edited by A. Briggs and W. Arnold. Plenum Press, 1996.

33. Fink M. Time-reversed acoustics. *Physics Today* 1997; 50: 34–40.

34. Stein L. Reciprocal action of reward and punishment mechanisms. The Role of Pleasure in Behavior: Harper & Row, 1964: 113–139.

35. Stein L., Belluzzi J. D., Ritter S., Wise C. D. Self-stimulation reward pathways: norepinephrine versus dopamine. J. Psychiatr Res 1974; 11:115–124.

36. Bishop M. P., Elder S. T., Heath R. G. Attempted control of operant behavior in man with intracranial self-stimulation. The Role of Pleasure in Behavior: Harper & Row, 1964: 55–81.

37. Heath R. G. Pleasure response of human subjects to direct stimulation of the brain: physiologic and psychodynamic consideration. The Role of Pleasure in Behavior: Harper & Row, 1964: 219–243.

38. Dobelle W. H., Mladejovsky M. G., Girvin J. P. Artificial vision for the blind: electrical stimulation of visual cortex offers hope for a functional prosthesis. *Science*, Feb. 1, 1974; 183: 440–444.

39. Folkard C., Freshfield J. Medical phenomena: Earliest successful artificial eye (care of William H. Dobelle). *Guinness (Book of) World Records*, 2005: 20.

40. Mowbray S., Jannot M. Best of what's next; five technologies that will transform your world: holographic TVs, spray-on space suits, bionic eyes, plastic buildings and interactive roller coasters. The bionic eye: we see the future better than 20/20; researchers have already restored some sight to the blind; why not give them super vision? *Popular Science*, June 2005; 266, 6: 58–59.

41. Siegfried T. Laser spots cancer before it grows. *NewScientist*, April 2–8, 2005; 186, 2493: 14.

42. Weintraub A. Genentech's medicine man, CEO Arthur Levinson got the biotech pioneer off life support; will it finally deliver on its promises? Tumors need blood, and they have a devious way to get it. *BusinessWeek*, Oct. 6, 2003: 72–80.

43. Pearlstine N. The next big thing: the laser unclogging arteries in the operating room. *Time*, Sept. 8, 2003; 162, 10: 75–80.

44. Selim J. The bionic connection: will neural implants erase the boundary between the mind and computers? Are we already a lot closer to a mind-machine interface than we ever guessed? *Discover*, Nov. 2002; 23, 11: 48–51.

45. Kohn D. Food for thought: the most sophisticated brain implant yet brings us one giant step closer to mind-controlled machines in news and views, headlines: pitting viruses against cancer, spray-on homes, a mind-controlled robot, robot subs, ready to serve. *Popular Science*, May 2005; 266, 5: 29–32.

46. Carpenter M. B. Core Text of Neuroanatomy: The Williams & Wilkins Company, 1972: 180.

47. Mancini L. S. short note: A magnetic choke-saver might relieve choking. *Medical Hypotheses*, 1992; <u>38</u>: 349.

48. Parosky M., neurologist at Erie County Medical Center, Buffalo, NY; personal communication: late 2003.

49. Andres J. C., Director, Family Practice Residency Training Program, Niagara Falls Memorial Medical Center, Niagara Falls, NY; personal communication: early 2005.

50. Cahill L., McGaugh J. L. Modulation of memory storage. *Curr. Opin. Neurobiol.*, 1996; <u>6</u>: 237–242.

51. Clark K. B. Post-training electrical stimulation of vagal afferents with concomitant vagal efferent inactivation enhances memory storage processes in the rat. *Neurobiol. Learn. Mem.*, 1998; <u>70</u>: 364–373.

52. Muir H. Hope for portable MRI. *NewScientist*, April 9–15, 2005; <u>186</u>, 2494: 9.

53. Witchalls C. At last a scanner that can see it all. *NewScientist*, April 16–22, 2005; <u>186</u>, 2495: 25.

54. Hogan J., Fox B. Sony patent takes first step to real-life Matrix: "A technique known as transcranial magnetic stimulation can activate nerves by using rapidly changing magnetic fields to induce currents in brain tissue. However, magnetic fields cannot be finely focused on small groups of brain cells, whereas ultrasound could be." *NewScientist*, April 9–15, 2005; <u>186</u>, 2494: 10.

55. Vogel M. Big minds gather to think small, really small. *Buffalo News* 1998; Sat., Oct. 24: C-5.

56. Kurzweil R. The *Age of Spiritual Machines: When Computers Exceed Human Intelligence*. Penguin Books, 1999: 52, 80, 120, 124, 127–128, 205, 220–221, 279, 300, 307–308, 313–314.

57. Kurzweil R. Live forever. *Psychology Today* 2000; Feb.: 66–71.

58. Kurzweil R. The coming merging of mind and machine: the accelerating pace of technological progress means that our intelligent creations will soon eclipse us – and that their creations will eventually eclipse them. *Scientific American* 1999; <u>10</u>: 56–60.

59. Wingert P., Brant M. Reading your baby's mind; new research on infants finally begins to answer the question: what's going on in there? *Newsweek*, August 15, 2005: 32–39.

Introduction

This is a book about two different kinds of predictions:

(1) What happens when we die, and
(2) What will happen on Earth within the next 1 to 100+ years.

The two kinds of predictions are remarkably similar to each other. If you were to die today, you would probably transcend instantaneously to a realm of the universe where science and technology are virtually *maximal* and *optimal*. This zone is not Heaven but rather it is a finite region that encircles planet Earth and is intermediate in many ways (including spatial location) between Earth and infinite Heaven. It is where all "deceased" and all yet-to-be-born earthlings' *minds* (or, synonymously, *souls*) reside and work on behalf of all currently "alive" (i.e., *incarnated*) earthlings. In this finite, intermediate zone which we can designate as either "Purgatory" or "Limbic Heaven," you would dwell in near-perfect harmony and work as part of a team with all of the other "deceased" and yet-to-be-born earthlings.

In Purgatory or Limbic Heaven, everyone is (1) famous (i.e., knows everyone else who is there *well*), (2) rich (i.e., goods and services are so plentiful that there is no need for money because effort and tediousness have been removed from learning and "work" via pleasurable mind/soul stimulation), (3) brilliantly knowledgeable to the point of near omniscience, for the same reason as (2), and (4) is otherwise fortunate for similar high-technological reasons.

If you were to remain alive here on Earth for the next 20 to 100+ years, you would witness the progressive advancement of science and technology toward the very same set of maximal and optimal conditions as referred to above. Additionally, through brain stimulation, reliably reproducible out-of-body experiences, global mind circulation and other technologies, you would be(come) nearly to fully liberated from your human body.

And you would come to resent having to spend *any* time inside of your body because you would come to realize that it is nothing but a punitive burden mandated by the cosmic envy (CE) of the infinite number of actual Heavenites (body-free mind/souls) who resent and are not yet ready to admit to us that they hate us for being residents of (center-stage-to-the infinite-universe) planet Earth and are entertained by our pain and suffering. If all of this seems like a load of wishful thinking, please keep in mind that the universe, being a place that runs on the pleasure principle, is also a place where wishes, more or less, dictate reality.

The Myth of Intelligence, Plus Everyone's Right to Painless, Suffering-free, Physician-Assisted Suicide

The human mind and the human brain are two different things. Your mind is a unitary physical particle. Its size is infinitely variable. It is potentially infinitely large. But depending on environmental circumstances, it can and does intermittently drop to zero size and thereby, temporarily, becomes nonexistent.

Its shape, depending primarily on your desire and choice, is also infinitely diversifiable. It may be referred to as a **mind particle** (**MP**), a **soul particle** (**SP**) or a **mind-soul-particle** (**MSP**).

You are your mind. So, you are your MP. You are not your brain. Your MP is *physically* bonded to your brain (and the rest of your body, too). But it does not require your brain (or any other part of your body) in order for you to be fully conscious and alert.

Your MP contains mostly only *conscious* (together with some subconscious) knowledge. Your brain contains mostly only *unconscious* (along with some subconscious) knowledge.

By means of the following phenomena:

1. electromagnetism (EM), possibly including lasers;
2. acoustic/sonic (sound) modalities (AS), possibly including ultrasound, light-reflective mirrors and multiple coil configurations;
3. chemical/pharmacologic agents (CPAs);
4. biometrics (BMs) such as brain waves (EEG), functional magnetic resonance imaging (fMRI), and optical tracking of eye movement patterns (EMPs);
5. biofeedback (BFK);
6. pleasurable brain stimulation (PBS);
7. pain/suffering-alleviative brain stimulation (PSABS);
8. reliably-reproducible out-of-body or so-called "near-death" experiences (ROBES or RNDES);
9. the Internet/World Wide Web (IWWW);
10. the Global Positioning System (GPS);
11. mind particle (MP) accelerators (MPAs) and decelerators (MPDs); and
12. instantaneous, global MP circulators and circulation (GMPCs and GMPC),it might soon become possible for every human being on Earth to have full, instantaneous access to *each* and (*nearly simultaneously*) *every* other human being's (as well as, of course, all of their own) conscious, subconscious and unconscious knowledge, emotions, thoughts and motivational states.

Sooner or later, most people will agree that there will no longer be any need or even desire for privacy, because everyone will benefit by being able to see and equally appreciate everyone else's viewpoint (and therefore be willing to compromise on almost anything and everything).

Plus, everyone will effectively *own* everyone's brain, body and life circumstances. So there will be few, little or no feelings of inferiority (hence, no jealousy), crime, terrorism or warfare.

Those traditional, old-fashioned-thinking individuals who continue to feel threatened by a blurring, blending, softening and equilibration of individual ego boundaries and, therefore, a continuing need or desire for a perseveration of the possibility of privacy, might be afforded the option of simply excluding themselves from participation in global MP circulation. However, this self-exclusion would entail forfeiture of huge amounts of knowledge, fun, friendship and love.

Consequently, everyone might soon be equally globally knowledgeable, skillful and peaceably, negotiably broad-minded, hence, equally well educated, capable and wise. So, everyone might soon be, effectively, for all practical purposes, *equally* intelligent in such ways as to relegate traditional ideas about inter-individual differences in innate intelligence, creativity, degrees of so-called "genius," practical skillfulness, etc., to the realms of mythology or, at most, ancient history.

The passageways toward this *egalitarian* endpoint of scientific, technological and philosophical progress might be expected to open up at irregular intervals of achievement/advancement, setbacks and clear-cut stages/milestones of enlightenment, spread out between *the present moment in 2007* and the general neighborhood of the end of this twenty-first century and beginning of the twenty-second century (i.e., the vicinity of 2099/2100) or shortly (within several decades) afterward.

The first major breakthrough (probably involving BMs, BFK, PBS and/or PSABS) might be expected to occur at any time between the present moment (now/today, 2007) and approximately thirty years from now.

This breakthrough might enable most of us to learn new information, concepts and work skills at rates, competency levels and in quantities that might be fifty to one hundred times (plus or minus) as fast, high and large as those we are currently able to do, that is, prior to any such breakthrough.

Also within the next thirty years or so, via BMs it might become possible, in a standardized way, to relatively objectively measure, categorize and quantify *pleasure, happiness* (contentment and joy), *pain* and *suffering*. Consequently, regardless of the plausibility of traditional ideas and arguments against painless, suffering-less and effective physician-assisted suicide, those notions may become widely discredited and legally discarded, so as to minimize or fully relieve needless pain and suffering.

HOW LEARNING ABILITY MIGHT BE IMPROVED
BY BRAIN STIMULATION

LEWIS MANCINI

25 Nottingham Terrace, Buffalo, New York 14216, USA.

Received: 28 November 1980

It would be necessary to determine an electroencephalographic characteristic, designated as the learning-linked characteristic (LLC), which always and only occurs during learning. The LLC would be used as a turn-on signal for an electric circuit which would deliver pleasurable electrical (or conceivably chemical) stimulation to a suitable reward site in the brain by way of an implanted intracranial electrode whenever and only whenever, and only for as long as the student would emit the LLC and hence engage in learning. From the student's standpoint, because learning and the LLC can only occur simultaneously, the learning process itself would be rendered pleasurable. And it would therefore be likely to occur for as long as the student's stimulation appetite remained unsatiated.

Hence the method, which could be referred to as learning facilitation or LF, would entail a prosthetic system consisting essentially of: (1) a *suitable* reward site in the human brain; (2) an intracerebral or scalp-level recording electrode; (3) an electric circuit which would recognize the LLC and use it to initiate and sustain stimulation of the reward site; and (4) an intracerebral stimulating electrode which would convey the stimulation to the reward site.

A practicable LF circuit would be similar to a circuit designed by Butler and Giaquinto[2] which delivered stimulation triggered automatically by electrophysiological events. They were interested in studying brain function in terms of the interplay between stimulated and unstimulated brain sites. Drugs could conceivably be used to maintain the thresholds of excitability of reward sites at low levels. In order to minimize the invasiveness of the system, the LLC would preferably be detectable at the level of the scalp rather than merely intracerebrally.

The workability of LF would depend upon: (a) the existence of one or more *suitable* reward sites in the human brain, and (b) the existence and detectability of one or more LLCs. The LLC might equally adequately be either unique to a particular individual whose learning is to be facilitated or common to that individual and many or all others.

A suitable reward site might be one which subserves pleasure which is inherently nondistracting to or integrable with the learning process. Using this kind of a reward site, the processes of LLC-detection, brain stimulation and learning could all occur simultaneously as implied above.

Alternatively, a suitable reward site might be one which subserves pleasure which is inherently distracting to learning but which also has the property that the pleasure does not appreciably outlast the duration of an electrical stimulus which induces it. If the pleasure significantly outlasted the stimulus duration, then, regardless of how brief the stimuli were made, the student would tend to waste time experiencing pleasure as opposed to learning. Because stimulation and learning could not occur simultaneously if this kind of a reward site were used, the electrical circuit would have to be designed in such a way that LLC-detection/stimulation periods would be alternated in time with learning periods. Both types of periods should be brief (for example, on the order of several minutes) because if the former were too long, the student would be wasting time experiencing pleasure as opposed to learning, and if the latter were too long, the student's learning process would effectively not have the benefit of pleasurable motivation.

An unsuitable reward site would be one which subserves pleasure which is inherently distracting to the learning process and which has the property that the pleasure does appreciably outlast the duration of a stimulus which induces it.

Although there is very little question that reward sites do exist in the human brain[1, 3], there would be little justification for hazarding any guesses with respect to their LF suitability. There is a wealth of experimental evidence that (at least relatively undetailed, slow-paced) learning can be facilitated in animals by means of rewarding brain stimulation[5, 9]. However none of this evidence seems to provide solid suggestions that the reward sites in question might be suitable according to the criteria given above.

Although at this time the question of whether or not the brain emits LLCs may not have a definitive answer, there are experimentally-derived suggestions that it does. Some of these are as follows. Lehmann[4] observed that "decrease of slow wave frequencies and increase of fast frequencies is systematically correlated with better quality of memory." Surwillo[7] found that longer as compared with shorter digit spans (as measured by the WISC digit span backward test) were accompanied by a greater degree of synchrony between the EEGs of the left and right hemispheres of the brain. Subsequently, Surwillo[8] found that tasks utilizing different cognitive modes have differences in their associated patterns of right-left interhemispheric symmetry of the EEG and that these differences, by means of the "central-moments technique of EEG period analysis," are detectable. Stigsby et al.[6], confining their attention to the dominant hemispheres of ten normal subjects, compared EEG recordings during auditory rest, an auditory memory task, visual rest and a visual memory task. They observed various increases and decreases in the amplitudes and indices of alpha, theta and delta activity as recorded from the various regions (for example, frontal, temporal, occipital) of the scalp. They concluded that "the two different types of mental activity, i.e., memorization and mental relaxation, induced two different patterns of regional EEG changes." And even if the brain does not, in fact, emit any LLCs which are common to statistically significant percentages of individuals in experimental sample-sized groups, the LF method may, nonetheless, be workable by virtue of the existence of LLCs which are unique to each individual.

If the LF method should prove workable, it would enable the brain to learn more effectively, rapidly and voluminously by virtue of reliably, rapidly and sustainedly providing large amounts of pleasurable motivational energy which could be used to accomplish the work of performing analyses and forming memory traces for informational perceptions. Additional learning-expediting benefits might be obtained by using LLCs which are specifically associated with very high-speed, detailed learning. LF would also enable anyone to become interested in any subject matter which would otherwise fall outside of that person's natural sphere of interests. It would do so by providing enough pleasurable motivational energy to overcome the pain of boredom, impatience and anxiety.

References

1. Bishop, M. P., Elder, S. T. and Heath, R. G., Attempted control of operant behavior in man with intracranial self-stimulation, in *The Role of Pleasure in Behavior*, R. G. Heath (ed.), Harper & Row, New York (1964).
2. Butler, S. R. and Giaquinto, S., *Med. & Biol. Eng.*, 7, 329–331 (1969).
3. Heath, R. G., Pleasure response of human subjects to direct stimulation of the brain, physiologic and psychodynamic considerations, in *The Role of Pleasure in Behavior*, R. G. Heath (ed.), Harper & Row, New York (1964).
4. Lehmann, D., *Electroen. & Clin. Neurophysiol.*, 30, 270 (1971).
5. Major, R., and White, N., *Physiol. & Behav.*, 20, 723–733 (1978).
6. Stigsby, B., Risberg, J., and Ingvar, D. H., *Electroen. & Clin. Neurophysiol.*, 42, 665–675 (1977).
7. Surwillo, W. W., *Cortex*, 7, 246–253 (1971).
8. Surwillo, W. W., *Physiol. Psych.*, 4, 307–310 (1976).
9. Guyton, A. C., *Textbook of Medical Physiology*, Saunders, Philadelphia, 762–763 (1976).

Editorial Comment

These suggestions of Mancini depend almost solely upon the existence of his postulated learning-linked-characteristic (LLC). I cannot see that such a characteristic can exist in the brain because there appears to be no necessity for this in the EEG wave forms. These wave forms are simply some weighted average of all of the millions of action potential spikes which are occurring in the brain at any one time. The recognition that learning has occurred is not a conscious or even unconscious action because presumably, on the basis of current physiological knowledge, memory is distributed over broad volumes of nerve cells at synapses and consists of a permanent or semi-permanent modification of the condition of those synapses. The idea, however, is a compelling one but this particular idea must concentrate on this matter because reward centres, etc., are quite well known and obvious but do not really bear on the practicality of the scheme Mancini proposes, without having LLC.

Response to Editorial Comment

The Editor(s) is/are wrong about this. There is <u>now</u> an <u>enormous</u> amount of experimental evidence that LLC <u>does</u>, in fact, exist. Some of that evidence is presented below in the essay completed on or before September 7, 2000, which has the following <u>short</u> title: *How Everyone WILL BE Rich, Famous, Painless, Deathless, Well Educated, Sexually Liberated, Etc., Starting Sooner or Later.*

Plus, it is probably either possible (or <u>almost</u> already possible) at this time (Sept. 2008) to dramatically improve anyone's learning abilities and work skills entirely surgically-noninvasively via a combination of focused ultrasound and electromagnetic phenomena. Moreover, it is almost definitely possible <u>at this time</u> to dramatically improve almost anyone's learning abilities and work skills surgically-<u>invasively</u> using electrodes implanted within the brain's "pleasure centers" in a context of deep brain stimulation (DBS) coupled with biofeedback. Shallow or superficial brain stimulation might play some role, too. In any case, please see the learning diagram that is displayed with its caption directly below.

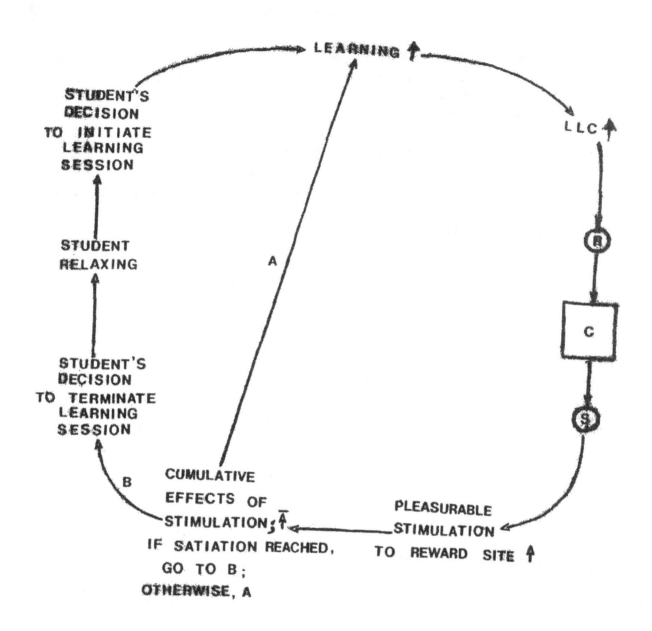

Figure One. Schematic diagram of how human (and animal) learning (and work skills performance) ability might be improved by a subtle variant or modified version of the psychologist, B. F. Skinner's operant conditioning (that is, essentially, a learning/educational-improvement paradigm that might entail instrumentally-mediated brain stimulation or some other kind of probably but not necessarily pleasurable and facilitative) mental activation. LLC = learning-linked characteristic, R = learned/newly-acquired response, C = central consciousizing circuit, S = (pleasurable) stimuli or stimulation.

Figures Two through Eleven (cartoons by Erica-Joani, circa 1990–1992): Schematic depictions of individuals enjoying and benefiting from learning-abilities-improving and/or work-skills-enhancing/ teaching-facilitative mental-sharpening equipment (such as, for example, brain stimulators) that also serve(s) as a mind (or, synonymously, a mind particle) circulator that might mediate interactions with other individuals' minds/mind particles, perhaps via satellite-mediated global positioning systems (GPS) or other kind(s) of mind (-particle) circulation-empowering devices or gadgetry, as conceptually explained in the longer of this book's two main essays.

Figure Two. A young man organizing his thoughts and putting his ideas down on paper.

Figure Three. Three men calmly sharing their thoughts with each other. They could be engaged in a process of highly significant and sensitive negotiations. Please note the relatively great distances between the contact-free brain stimulators/mental activating equipment and the three gentlemen on whom these remote neuroprostheses are respectively focused.

Figure Four. A woman and man calmly experiencing the benefits of couples' counseling.

Figure Five. Three athletes experiencing the joys of motivational enhancement.

Figure Six. An agitated, potentially violent individual being calmed down and somewhat relaxed so that he can deal with his problems in a relatively rational way.

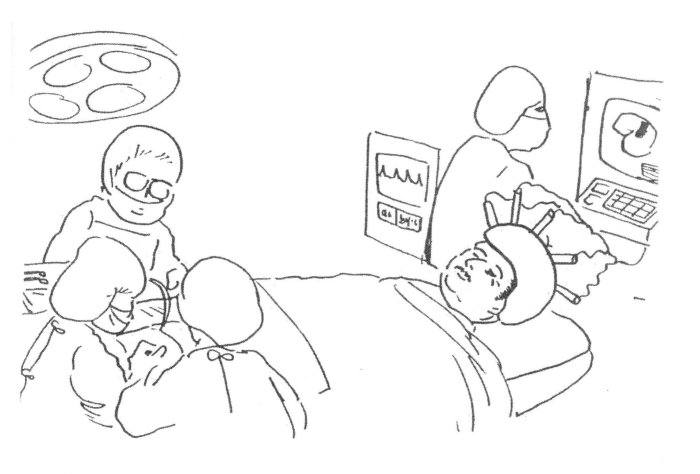

Figure Seven. Pleasurable-mental-activating/brain-stimulating painlessness/analgesia being enjoyed by a patient while he is undergoing surgery.

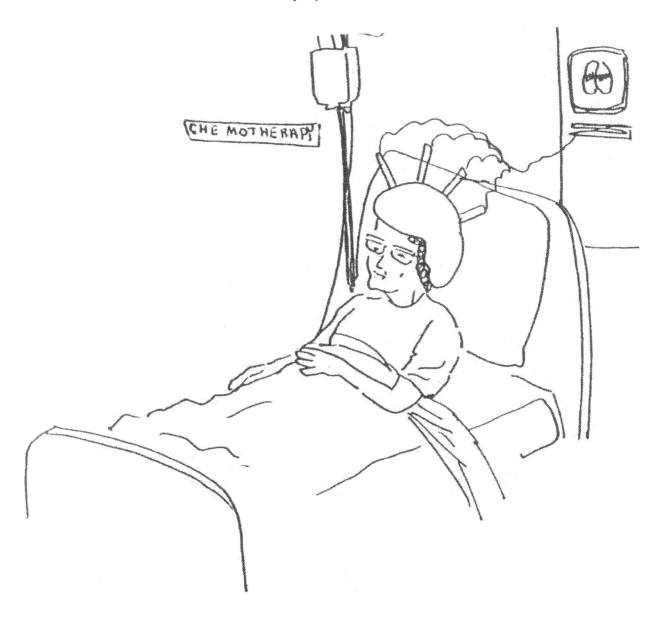

Figure Eight. Pleasurably-brain-stimulated, hedonic-calculus-maximized and, mentally-activated freedom from side effects associated with medical-therapeutic treatment(s) for an individual's illness/medical problems.

Figure Nine. A young man enjoying pleasurable brain stimulation or other possible mode(s) of rewarding mental activation in a context of learning facilitation or work-skills knowledge augmentation and diversification.

Figure Ten. A young woman enjoying learning facilitation and work skills enhancement and diversification.

Figure Eleven. A young man who is simultaneously deriving entertainment, educational, vocational/occupational and employment-related value and enhancement via rewarding brain stimulation/mental activation.

Preview of Material Written in 2008

Pleasure is the only reason why the universe or anything in it exists. Without pleasure, there would be nothing but infinite, empty space and eventless time. The universe is infinitely large and contains an infinite number of ordinarily indivisible, self-cohesive (but vanishingly soluble), unitary-identity-of-consciousness-containing, physical particles, each of which is capable of experiencing pleasure, pain, suffering and consciousness. Pain and suffering could not exist at all unless the ratio of pleasure to pain in the infinite universe at large were infinity to one (∞ : 1). Hence, every particle in the universe is a single mind (M) or, synonymously, a mind particle (MP) or, synonymously, a soul (S), soul particle (SP) or mind-soul-particle (MSP).

Because the universe is infinitely large and, thus, has no shape-defining outer boundaries, in *order for* any given MSP to communicate with and specify a point in space to any other MSP(s) (such as in, "Let's get together soon, maybe a hundred years from now, at point x = 10^6 (one million), y = 2 x 10^6 (two million), z = 3 x 10^6 (three million), so that we can collaborate on that project we've been talking about"), there *must be* a *cosmic central-reference-point* or *focal point planet* (CCRP) (or simply CRP or CFP) within the universe whose centerpoint or "origin" has the geometric coordinates: x = 0, y = 0 and z = 0. Let us consider the hypothesis that planet Earth (PE) is the current CRP point of the spatially-infinite universe. Consequently, each MSP trapped inside a human body (each earthling) is *already* (unbeknownst to herself/himself) *infinitely famous*, because an infinite number of *non-Earth-bound* MSPs have their envious attention riveted on the interplay among all of us earthlings. If this were true, then William Shakespeare might have been a physicist as well as a playwright when he wrote "All the world's a stage and all the men and women are merely players …" in his play *As You Like It*. Hence PE might be to the universe what television (and other media) are to us. Perhaps the infinitely large audience of non-Earth-grounded MSPs are *entertained* by all of our pain and suffering! *In order to prevent* PE (the current CRP) from being chaotically inundated, overwhelmed and ruined as such (i.e., a centerpoint for the universe) by infinite hordes of jealous, attention-seeking, narcissistic MSPs who all want to be on "center-stage" (like a theatre stage being overrun by throngs of irate, unemployed actors), the PE/CRP *must* be riddled with pain and suffering, which serve as an impediment, deterrent, deflecting force or even an insurmountable barrier to the infinite number of would-be invaders. Even so, some MSPs might be able to trickle through or terrorize PE/CRP in the form of (often havoc-causing) "ghost particles," neutrinos, cosmic rays, "flying saucers-?" etc.

So, the *infinite* number of MSPs who are not of PE (whom we might plausibly think of as "space aliens") are *envious* of the finite number of us MSPs who do live on this planet. Consequently, due to their *cosmic envy* (CE) of our centrality to the universe, by invisibly but vindictively trickling through our world as the particles they are or by transmitting hostile waveforms through our homeland, they create *additional* pain and suffering for us, in the form of physical and mental illness, financial stress, learning disabilities, natural disasters, etc. Hence, being grounded on Earth as we humans are, is, at best, a dubious question of being "lucky enough" or lucky at all.

With each successive Big Bang *Re*-creation of our cyclical universe, there comes a new choice of CRP and a new initial choice of a finite number of MSPs, who will ever be allowed to live there, through one or more lifetimes within the new central planet's prehistory and/or history. This finite number of chosen MSPs can be considered tantamount to the royal family of the universe, not only because of the center-stage privilege, but also because of the glorious destiny (beyond the scope of this preview) that awaits them.

When a CRP is newly designated as such, CE is maximal and, therefore, conditions on the planet are maximally Hellishly brutal (that is, full of ignorance, hatred, hardship and violence). Painfully, slowly and gradually, CE subsides over billions and millions of years and begins to rapidly subside over thousands and hundreds and decades of years. So, the (that is, in the current case of PE/CRP, *We*) inhabitants of the central planet are begrudgingly given progressively more and more scientific and technological enlightenment. What we call "progress" is simply the result or reflection of the gradual diminution over time of CE, as the infinite number of CRP-excluded (hence, hostile) alien-to-CRP-space MSPs mellow, "get over" and adjust to the fact that they did not get included in the most recent Big Bang Family (BBF) membership list of cosmically central MSPs.

What progress means for us contemporary Earth-dwellers in terms of the relatively near future (perhaps the next hundred years or so) might be as follows: Each person is an individual, irreplaceable and unique "building-block" of the infinite universe, who is ageless and *immortal*, but temporarily trapped inside of a *multi-component*, hence ageing (i.e., falling apart into individual component parts) and, therefore, mortal body.

An extrapolation of the world's getting progressively "smaller" as time goes on (with humans living in isolated caves long ago to our now being able to contact anyone on the planet instantaneously) will lead to a condition, probably sometime within the next hundred years, whereby every person (i.e., MSP) who has ever been or ever will be born on Earth will have direct access (via the Internet/www, GPS, reproducible out-of-body experiences, global MSP circulation, etc.) *to the identity* (hence, *fame*), the *finances* (hence, *fortune* or *wealth*) and the thoughts/*knowledge*/emotions (hence, *brainpower*, friendship and love) of every other person who has ever lived on this planet.

And, due to the de-effortization of learning and "working" via brain and MSP stimulation (together with some other technologies), the total knowledge and wealth of the world will be virtually infinite, not only for all of us, collectively, but also for each and every one of us, individually. Hence, what Carl Jung referred to as the "Collective Unconscious"(-ness) will become a full, global Collective Consciousness, without any loss or impairment of each individual MSP's *unique* identity.

The reason why each person/mind particle is trapped inside a problematic, burdensome human body has nothing to do with evolution (which is true and valid, nevertheless) and much to do with an uncanny, highly abstract form of creationism/intelligent design, as well as the *hostile* force of CE on the part of that infinite number of embittered and resentful mind/soul particles who were excluded from our center-stage world at the time of the most recent Big Bang explosion and associated CRP selection.

And the astrophysically observable "reality" that planet Earth is an average, ordinary planet is actually just an illusion foisted upon us by that same infinite angry load of Big-Bang-excluded souls who cannot tolerate the thought of our *both*: a) <u>being central</u> to the universe, and b) our "smugly" *being aware* of our cosmic centrality (CC).

Moreover, despite the privileges of CC, because living on Earth entails a burden of pain and suffering that probably exists nowhere else in the infinite universe, each and every one of us earthlings (earthbound MSPs) should be acknowledged to have the right (and granted access) to painless, suffering-free, physician-assisted suicide, not quite "on demand," but certainly whenever reasonable criteria of a poor quality of life have been met.

In line with this last point is the observation that it never ceases to amaze me how human MSPs, or at least the legal system that is supposed to represent us dictates that the merciful route of euthanasia is good enough for the MSPs of our beloved pet dogs, cats, horses, etc. (humans are hardly the only MSPs that are trapped inside of and burdened with bodies), but euthanasia or painless, suffering-free, physician-assisted suicide is apparently *too good* for us human MSPs when *we* are suffering.

In other words, the legal prohibition against virtually all forms of human euthanasia, together with the tolerance for subhuman species' euthanasia, amounts to treating animals humanely while treating humans inhumanely. Hence, when animals suffer, we take the attitude that "Euthanasia is the only humane thing to do," but when human beings suffer, we take the attitude that "They'll just have to live with *their* suffering." This paradox seems *most unfair* when viewed from a compassionate standpoint. The range of options for dealing with human suffering (*including profound unhappiness associated with mental illness*) should be at least equally broadly inclusive as that for dealing with animal suffering.

EQUAL FAME, ETC.

Including Financial Security, Brainpower, Euthanasia and Immortality for Everyone, Forever

Chapter One
The Necessary Existence of Purgatory
As Well As Heaven and Hell

The idea that planet Earth may be the Hell-centerpoint of the infinite universe may seem both (a) too pessimistic and (b) too grandiose to be true. Regarding (b): the grandiosity of what is being proposed herein is precisely the aspect of our situation which the universe at large (i.e., Heaven) begrudges letting us take due pride in.

Regarding (a) the pessimism of what is being proposed herein: it does not seem so pessimistic when you consider that the Hellishness of the human/earthly condition is probably nearly over with. If our Hell was born approximately 15 billion years ago, at the time of the most recent (of an infinite number of consecutive, cyclical) Big Bang Creative Explosion(s), and if this Hell will be over with by the year 2100, then our Hell is approximately 99.99999999387% over with! That doesn't sound so pessimistically bad!

Let's pretend that we have a time machine (i.e., a real possibility) in our hands at this time and we choose to jump ahead to the year 2100. OK. Every human being on the planet now knows every other human well. Hence, everyone is world-famous. And every human has been reunited with all of their "dead" relatives and friends (i.e., their "dead" relatives' and friends' immortal mind-soul-particle(s)/MSPs). Hence, no love has been lost, after all and despite all of the mind games that the infinite number of hostile-to-us Heavenites love to play at our expense!

Thanks to brain stimulation and mind particle circulation, everyone has become so knowledgeably "intelligent" and so capably productive (effortlessly so) that goods and services have become so inexpensive that money has become obsolete. And, consequently, everyone is now effectively a trillionaire by the standards of 92 years go (2008). Hence, everything is wonderful to the degree that Earth no longer qualifies as Hell, but has been fully Heavenized and has become one of and merely blends in among an infinite number of wonderfully Heavenly planets that anyone can spend time on.

Thus, we have the joyful experience of a thoroughly wonderful, infinitely large, Hell-free universe. But, the state of Hell-lessness, as wonderful as it sounds, is actually a deeply problematic state that will engender the next Big Bang Explosive Collision and the birth of a new Hell and a new Hell-to-Heaven cycle of prehistoric deep Hell that will slowly evolve into another Heavenly planet.

A universe without an attention-grabbing Hell-centerpoint-planet is a universe without a clock by which to measure absolute time as well as without a spatial central-reference-point by which to measure, quantify and specify space whereby joyful events in Heaven could be planned and orchestrated. In order to focus the infinite universe's attention on one specific planet, a new Hell (whereon entertainingly Hellish lives are led) MUST be born. Hence, 2100 might mark the end of Earth-Hell, but no sooner will our Hell be over with than a new Big Bang Hell-creation will be in progress.

You and I and all other graduates of (Earth-) Hell will be big celebrities in Heaven at large, wherein for every Hell-graduate there are and will probably always be an infinite number of MSPs who have never been to Hell, at least not as full-fledged residents. And this infinite number (compared to each one of us Hell-graduates) will, figuratively speaking, be after us for our autographs, celebrate us as one-in-an-infinity celebrities, etc.

So, the bad part of being a Hellite is *while* we're here. After we leave (both Purgatory and) Hell, we'll be super-privileged, super-celebrities for the rest of infinite time. We'll all be true "big shots" after we leave Hell and Purgatory as one big happily-relieved Big Bang (virtually Royal) Family. We'll be virtually worshipped and actually revered as graduates of Hell, spreading out among the infinite hordes of MSPs who have never been unlucky/lucky enough to live in Hell-Purgatory.

Here is how the next Hell might develop. There are an infinite number of MSPs floating around in infinite space, all basically having a frolicsome time (frolicsome except for their envy of post-Hellite-Purgatites). Each MSP in Heaven, whether an "ordinary" MSP or a post-Hellite-Purgatite MSP, floats in and out of full conscious awareness/awakeness and diffuse, sleepy, subconscious or unconscious awareness, much as we alternated between being awake and asleep back in 2008.

Now, we're "dead" and have no body/ies to bother us. So each of us is just a particle (which can assume any size, shape and appearance we want) that flickers in and out of existence just as we flicker back and forth between full consciousness and full unconsciousness. When we're subconscious, we still exist as particles but our size is small and our internal particle substance is sluggishly stagnant as opposed to flowing around freely and vibrantly when we're fully conscious.

Although we're all having a wonderful time now in 2100, we realize we need a new centerpoint-Hell by which to gauge time and space so that we can agree on when and where to meet for any collaborative occasion, such as, for example, our Big Bang Royal Family one-year reunion in 2101. So, how do we all (all infinitely many of us throughout the infinite universe) go about generating and designating a new Hell?

Well, we advise each other to "just go with the flow" and move in whichever direction gravity seems to pull us. In this way we all gravitate toward some one MSP in much the same way as most of us Americans gravitated toward the presidential and vice-presidential candidates in the U.S.A. in 2008. And similarly, as we finally elected/selected Barack Obama and Joe Biden to lead us in 2009, we (the infinite number of MSPs) will elect/select some MSP (not necessarily superior to any other MSP and we're not talking about God here) to select and engineer the construction of some planet which will become the next cosmic central-reference-point for the infinite universe. Let's not incorrectly call this elected/selected MSP "God." Let's just call them/her-him (neither she nor he, because gender is merely a contrivance of Hell), the "President of the Universe" (POTU).

Once the infinite number of us MSPs have selected a new POTU, he/she will subsequently, decisively announce: "And the next central-reference-point Hell will be right HERE," as she/he seemingly, almost randomly points to/designates some spot in the universe (actually a point of relatively maximal gravitational attraction) where a planet will materialize (made up of many, many subconscious MSPs that are bonded to each other by strong gravitational and other attractive forces) and toward which all MSPs who have never lived in a Hell, competitively race in hope of being included hereon as full-fledged residents.

All of us MSPs who have lived in Hell know better than to try to become doubly renowned by living in two consecutive Hells, which is probably against cosmic-universal rules anyway, with the possible exception of the POTU for whom the compassion gained by having already lived in a Hell may be a prerequisite for becoming the POTU. But there are always infinitely more MSPs who have never lived in a Hell than there are MSPs who have ever lived in one. You and I will probably avoid going to Hell twice for obvious reasons, even apart from the rules.

However, a finite number (let's say 10 raised to the power of 10,000 or one followed by 10,000 zeroes) of MSPs, let's dub this number, loosely speaking of course, a "zillion" never-lived-in-a-Hell (NELINIAH) MSPs will race toward that next central-reference-point-Hell (CRPH), in effect, shouting, "Get out of my way! I want that center-stage Hell! You can just wait on the sidelines in Purgatory!

"Give me that already-ready part in the Hell-play! Let it be mine! Now! You can wait till some other part in the Hell-drama-comedy-tragedy comes up for grabs. You can just wait behind the scenes in Purgatory, while helping those of us out who are lucky enough to get already-ready parts in the play (such as those of dinosaurs, etc.)!"

These zillion NELINIAH MSPs will all race toward that infinitesimally small point where the CRPH planet is about to materialize (made up of some of the most impatient MSPs). And they will be rushing so madly, these greedy-for-fame-crazed NELINIAH MSPs, that they will all collide so violently with each other that there will be a BIG BANG sound emitted on impact.

And with this BIG BANG implosion into a tiny sphere of space, a new CRPH will be born and all zillion NELINIAH MSPs included therein will collectively constitute the new BIG-BANG-ROYAL-FAMILY (BBRF) of the universe. Whether they get an immediately-ready part in the Hell-play or have to wait for later-on-developed parts, they'll all be either center-stage or just-outside-of-the-focal-point (of the Hell, i.e., in the Purgatory) of the universe. The Big Bang phenomenon is at first an implosion (i.e., collision), but then rebounds or ricochets into a colossal, expanding explosion. So the Big Bang phenomenon may be referred to as either an implosion or an explosion (or both).

Actually, many zillions of MSPs will race toward the centerpoint, but the POTU, who is allowed to grow to an enormously powerful size, will decide which zillion actually "makes the cutoff" and is/are each allowed to participate in the actual Big Bang. However all of the many zillions, indeed an infinite number of MSPs who didn't make the cutoff (imposed by the POTU) will be so enviously angry that they will passionately hate all of the MSPs who did make it.

Hence, even in Heaven, everything is not perfect. Heaven is imperfect because you have an infinite number of NELINIAH MSPs who are crazed with jealousy (i.e., COSMIC ENVY, CE; could this CE be akin to or even equate with "original sin"?). They are crazed with jealousy, CE, because when the POTU said, "OK, the rest of you will just have to hold back, because I have enough aspiring actors and workers for all the parts in the Hell-play and for all of the jobs in the Purgatory-Work-place.

"I'm sorry you cannot be involved in this BIG BANG collision (and, therefore, you're not part of the new BBRF, but better luck next time around the BIG BANG cycle of birth of a new Hell which will, over 15 billion years or so, progress back toward a state of de-Hellification and re-Heavenization. So, just back off for now."

The POTU will have been given so much power by the universe that when S(He) says you must stop and back off, then you must do so. But the infinite number of NELINIAH MSPs who have been precluded/excluded by the POTU from the BBRF will be so bitter and furious that they will make life for all of the BBRF members a living Hell!

OK, now in the year 2100, we have this brand-new Hell and a Heaven full of furiously jealous NELINIAH MSPs who won't tolerate the BBRF being center-stage unless they are stripped of almost all knowledge and benefit of convenience of science and technology and forced to live in caves, hunt down dinner in order not to suffer the misery of starvation, endure all kinds of diseases, etc.

So, what is Purgatory and why do we need to have it? OK, the POTU informs all of the BBRF member MSPs that they cannot all simultaneously be center-stage. Some of them need to negotiate peace and progress and serve as a buffer zone (i.e., Purgatory) between angry Heaven and paradoxically privileged-yet-victimized Hell.

The POTU needs assistants in much the same way as the President of the United States needs his/her vice president, Cabinet, Congress and the Supreme Court. And this assistance will be needed for approximately 15 billion years subsequent to the Big Bang Explosion that occurs after each consecutive Hell becomes fully de-Hellified and Re-Heavenized.

Let's hypothesize that of the zillion MSPs who become BBRF members in 2100, only perhaps 200 billion will get assigned to human roles (will get incarnated into human bodies) at various times throughout the 15-billion-year BIG BANG Cycle (BBC), mostly during the last two million years or so of the cycle. The other zillion MSPs minus 20 billion (MSPs) get incarnated into/assigned to subhuman (e.g., dinosaur), other kinds of animal, plant and inanimate object roles within the new Hell-centerpoint planet. So, even inanimate objects are composed of many subconscious souls (MSPs) and should be treated with respect.

Inanimate objects such as computers, chairs and floors (within Hell) are actually huge conglomerates or aggregates of subconscious MSPs which work together to maintain the structural and functional integrity of the "inanimate" object. So, for example, a computer is actually a large number of subconsciously MSPs working together in virtually perfect harmony.

And each of the MSPs in question is a BBRF member in its own right. Our bodies are also made up of a huge number of BBRF member MSPs who are working together subconsciously. And the dichotomy between animate and "inanimate" objects is actually a false one. Remember: All particles are MSPs, which are merely the elemental building block particle(s) out of which everything and everyone is made. The only difference between an animate and an inanimate MSP is that an animate MSP, such as you or I, is a generally fully conscious one (except, for example, when we're asleep), whereas an inanimate MSP, such as all of those which compose the bodies that we've been incarnated (by the universe, our Heavenly enemies) into, are subconscious at virtually all times.

Hence, the vast majority of the zillion BBRF members will never get a human role in the play on the new Hell-centerpoint-planet. And of the 200 billion or so who will get a human incarnation, not all will get it at the same time. On the contrary, all of the human roles will be spread out in time over a two-million-year period or thereabouts.

The POTU and her/his assistant BBRF MSPs (Purgatites who are either waiting for or who have already had/lived and died from their Hell-parts-roles-in-the Hell-play/movie) will decide on when and in what specific context and interrelational role each of the approximately 200 billion human-incarnated MSPs gets to play her or his role in the Hell-centerpoint play.

All of the MSPs who are destined to have human incarnations but who reside in Purgatory (the buffer alone between hateful Heaven and unsuspecting Hell) will help the POTU to negotiate peace between Heaven and Hell, including the gradual restoration of all of the science and technology (progress) that the infinite number of hateful NELINIAH Heavenites took away from the Hellites in their anger after the Big Bang that occurred most recently circa fifteen billion years ago.

Hence, all of the Purgatites, including human-designated MSPs, who have already lived and died in the new Hell, as well as those whose roles have not yet begun, will serve as advocates for all of the human-incarnated (and all subhuman) Hell-bound-resident-MSPs.

So, Purgatory is necessary as a place where the POTU and her/his many, many non-Hell-resident assistants work to negotiate peace and progress that will take place and eventually de-Hellify and re-Heavenize the new Hell that will be born around the year 2100.

In summary, Heaven is an initial, infinite place, where particles pop in and out of existence and operate strictly on the pleasure principle. Hell is inevitable because the universe, that is, Heaven needs a central-reference-point by which to measure absolute time and space and also because the Heavenites get tired, weary and jaded in relation to everything and everyone being so pleasant all the time and actually want to see and be entertained by some pain and suffering (which they inflict on the Hellites).

Other reasons why Hell exists are because all Heavenly MSPs (i.e., Hellites) crave to be universally famous as well as to become members of what inevitably become internally loving and egalitarian Big-Bang-Royal-Families.

Purgatory is necessary because there must be an out-of-focus periphery of the central-reference-point. Purgatory serves as a buffer zone between Heaven and Hell and from which the POTU and his/her many assistant MSPs can successfully beg for mercy from and gradually appease the angry NELINIAH Heavenites in relation to the being-kept-in-the-dark, unsuspectingly victimized Hellites who are actually, unknown to themselves, already big celebrities within the infinite universe (and are allowed/will come to know of their/our own celebrity during their/our Purgatistic stints which occur before and after their/our stints in Hell proper.)

Why Hell-bound MSPs are not allowed to know of their/our cosmic-infinite-universal fame is just part of the "hell" of being in Hell. The angry Heavenites want us victimized Hellites to feel relatively worthless, insignificant and anonymous, despite the reality that nothing could be farther from the truth within the infinite universe. Most people are paradoxically unfamous throughout the world/Hell, but all people are infinitely famous throughout the infinite universe.

In any case, we need Purgatory because we need a large number of advocates (MSPs serving as advocates) who will work with the POTU to bail us all (the entire Purgatory-Hell-bound current BBRF) out from our current (Purgatistic and Hellish) conditions and circumstances.

Saying that the POTU needs Purgatites to help him/her negotiate with the angry Heavenites on our behalf is a little bit like saying that if there were a Santa Claus, he would need some helpers (i.e., the elves), which stands to reason, given the hypothetical enormity of Christmas holiday tasks. And saying that we Hellites need to have Purgatites working (imperceptibly from our viewpoint) on our behalf is like saying when and if you have profound legal problems, you might need more than just one lawyer, one expert, one character reference, etc.

In summary, we (finite number of) Hellites need at least a (finite) number of friendly (but invisible and otherwise imperceptible) Purgatites to help us in our struggle against the infinite number of hostile (and imperceptible) Heavenites, who are incessantly harming us (sending poverty, disease and ignorance our way) via waveforms and MSPs that travel through us via imperceptible dimensions of space (of which there are an infinite number).

It makes sense to pray to and for the Purgatites, who are our "dead" loved ones and friends and their loved ones and friends and all other members of our Big Bang Royal Family, including those members who have yet to be born (in any human or subhuman capacity) here in Hell.

Our Purgatite friends send us good things (money, good health, knowledge, etc.) via imperceptible waveforms and MSPs that shape and sometimes change our circumstances here in Hell for the better. By the year 2100, the POTU, working together with our Purgatite friends, will probably have succeeded in fully de-Hellifying and Heavenizing our world.

Hence, we do need Purgatory as a finite, peripheral place (intermediate in spatial location between finite Earth-Hell and infinite Heaven) wherefrom our POTU and our Purgatite friends can help us in our struggle against our infinite number of NELINIAH Heavenly enemies.

Moreover, the observation (by astronomers, astrophysicists, cosmologists, etc.) that Earth is just an unremarkable planet in an unremarkable galaxy in an infinite expanse of galaxies is true in a way and false in a way. It's true in a physical sense, but false in the sense that planet Earth is the arbitrarily selected chronological and spatial central-reference-point within the universe.

Our jealous Heavenly enemies want (!) us to incorrectly think that we have no special role in the universe. But a world as troubled and problem-fraught as our world is / would be too unstable to exist at all without the imperceptible help of the POTU and the Purgatites and would also be too unstable to exist unless our Heavenly enemies wanted us to exist as such for all of the following reasons:

1. They need a spatial reference point in order to work collaboratively among themselves;
2. They need a chronological reference point in order to work collaboratively among themselves;
3. They get tired of everything being so pleasant in Heaven and want to be entertained by our pain and suffering;
4. They want to be famous throughout the infinite universe as we Hellites already are;
5. They want to be members of an internally loving, egalitarian BBRF the way all of us worldly Purgatites and Hellites will be after full de-Hellification and Heavenization occurs around the year 2100, and
6. They want to become major celebrities throughout Heaven as a consequence of becoming celebrities on account of having lived in and graduated from a Hell and a Purgatory the way all of us Earthly Hellites and Purgatites will be after approximately the year 2100.

Thus, Cosmic Envy (CE) consists not only in the infinite number of NELINIAH Heavenites being currently jealous of us Hellites and, less focusedly, our respective Purgatites on account of our being center-stage to the infinite universe, but also in their envious anticipation of the huge pain-and-suffering-free celebrities that all of us Heavenites/Purgatites will become after we graduate from Hell and Purgatory.

Hence, in a way, Copernicus, Galileo, Einstein and all the other luminaries of physics have missed and continue to miss the intuitively inferable realization that Earth is functionally the center of the infinite universe, despite the fact that in a geometric, spatial sense, an infinite volume of space (i.e., the universe) cannot possibly have a true center.

Moreover, new suggestive evidence for "the anti-Copernican idea that we're special after all—that our visible universe is an unusually low-density region of a much larger cosmos" is beginning to emerge (ref. 1).

Reference

1. Panek, R. Going Over the Dark Side. *Sky & Telescope*, Feb. 2009; 117 (2): 22–27.

Chapter Two
Is Jesus Christ the President of the Universe (POTU)?
(Plus, more on Hell, Purgatory and Heaven)

Is Jesus of Nazareth merely the current president of the universe (POTU) who was elected by the universe at the time of the most recent Big Bang Explosion that occurred some 15 billion years ago, which would mean he is no more intrinsically superior to you or me than President George W. Bush or President-elect Barack Obama is intrinsically superior to us (or anyone else)?

Or are Jesus (and the other two parts of the Trinity), taken together, one God who always has been and always will be intrinsically superior to the POTU and all of the rest of us humans as well? This question, in terms of a detailed analysis, is beyond the scope of this book, but speaking as the son of my Catholic father, I would say "Yes" to this. Speaking as the son of my atheistic mother, I would say "My opinion on the subject is a complicated one that is beyond the scope of this book."

If you and I and all of the rest of us Hellites could see how wonderful our immortal lives will be after we graduate from Hell/Purgatory (the "dead" and the yet-to-be-born can already see it, because they're in Purgatory which is not nearly as bad as our Hell-Earth), if we could see what grand pain- and suffering-free immortal celebrities we'll all be after we graduate from Hell/Purgatory, many or most of our "death wishes" would be much stronger than they currently are. In fact, many more people would be clamoring for voluntary, painless, physician-assisted suicide than currently are doing so. The Purgatites' lives are not nearly as pleasureful as those of the hateful (NELINIAH) Heavenites! The Purgatites' lives, though pain and suffering free, are relatively bland pleasure-wise, yet gratifying in terms of helping their fellow BBRF members who are still trapped here in Earth-Hell (i.e., all human and subhuman earthlings, including us "living" humans).

Plus: "The universe is getting more complicated. In addition to gravity, dark matter and dark energy, we now have to deal with dark flow. This is a sideways movement of thousands of galaxy clusters (that) seems to extend across the entire sky. The most reasonable explanation (for this) is that there is an <u>attractor</u> somewhere farther away than the limits of the observable universe (ref. 1)." The observable universe might include Earth, Purgatory and some finite volume of Heaven. The "attractor" could be the infinite expanse of Heaven pulling Earth and Purgatory back into it at an ever accelerating rate, so that Earth and Purgatory will be fully Heavenized/reabsorbed into Heaven by the year 2100 or thereabouts.

Reference

1. Marlow, D. Dark Flow. *Mensa Bulletin*, Jan. 2009; <u>521</u>: 49.

Chapter Three
Evolution versus Creationism versus Intelligent Design

All three concepts are valid. Evolution is valid both in the sense of literal Darwinian "survival of the fittest" and also in a more subtle sense of the universe (being led by the POTU) gradually progressing through successive kinds of phases of conception much as an artist might move through his/her blue period, red period, green period, etc.

The universe creates us in the sense that even though we never existed for the first time (you're a particle that has flickered in and out of existence since forever, i.e., negative infinite time), we only take on an identity in the "eyes" of the infinite universe by dint of becoming an actor on Earth-Hell's stage.

Before you came to Hell during the last Big Bang Explosion/Collision some 15 billion years ago, you were an unknown never-lived-in-a-Hell (NELINIAH) Heavenite who always aspired to a role as an actor in Hell.

So, when you joined the cast of Earth-Hell's play 15 billion years ago, you were created, which is to say, you were "discovered" in much the same way as an "unknown, ordinary" person might suddenly be created/discovered as a movie star in Hollywood, California, or as a notable politician in Washington, D.C. In other words, you always existed, but you weren't created/discovered until you became a star on "Earth-Hell-TV" whereby you were in line to entertain the infinite universe at large (i.e., Heaven) with your pain and suffering.

Hence, the universe created Earth-Hell-TV because it needed a central chronological and spatial reference point. And the Earth-Hell created/"discovered" you by making you a movie/TV star here in Hell. And the reason why you're not being informed (by the universe) of your Hell-TV movie-star status is because of cosmic envy on the part of the always infinite number of still NELINIAH Heavenites, plus the fact that they would not be entertained by your pain, suffering and boredom if you realized there were a cosmically important and crucially necessary purpose for all of that bad stuff that does (and good stuff that doesn't) happen to you here in Hell.

Obviously all of the above constitutes Intelligent Design on the part of the universe and its elected POTU. As for mutually disconnected parallel universes and multiple universes (i.e., a "multiverse" [ref. 1]), that's probably not, strictly speaking, true. But please see further explanation that is directly following.

In an interconnected sense, the universe does have parallel and multiple aspects. Since the POTU and all of the rest of us MSPs are basically similar to each other, the physical nature of the universe follows from the psychodynamic aspects of the highest-functioning echelon, i.e., the MSP that is incarnated into a human body. Hence, the universe is anthropic in nature.

So, the nature of the human mind-brain pairing dictates the physical structure of the universe. And the universe is anthropic because human nature is such as to want, will and, consequently, bring about or engender a single, infinitely large, internally integrated/interconnected universe rather than an infinite array of mutually independent universes that might not even be aware of each other's existence.

However, since there are an infinite number of spatial dimensions, as discussed in another chapter, there can be and are mutually imperceptible phenomena that are parallel to each other. Hence, there are mutually parallel or multiple phenomena, but they're all integrated into the same, infinite, unitary, anthropic universe in the sense that they all have the same central spatial and chronological reference point (i.e., planet Earth-Hell), (along with its peripheral, surrounding Purgatory) and the same POTU.

Regarding the question of reincarnation, it is probably true that in rare instances, where an MSP-person (an MSP incarnated into a human body) has behaved unusually wickedly during his/her lifetime, an MSP can be and sometimes does get reincarnated via action taken by the POTU and the Purgatites (working together) and thereby gets sent back from Purgatory to withstand a second round of human life on planet Earth-Hell. But such occurrences are probably very rare and exceptional, fortunately.

Another relevant point is that all particles in the universe, whether they're animate or inanimate, are mind-soul-particles (MSPs). The only difference between an animate and an inanimate MSP is that an animate MSP has its consciousizer or consciousness mechanism temporarily turned ON (which means that its internal substance flows around quickly and briskly inside of it), as opposed to an inanimate MSP which has its consciousizer or consciousness mechanism temporarily turned OFF (which means that its internal substance flows around slowly and weakly inside of it).

Hence, any MSP is animate when it is fully conscious and inanimate when it is merely subconscious. When an MSP becomes temporarily unconscious, then it temporarily disappears from the universe. But since all MSPs are immortal, it (the disappeared MSP) will reappear whenever and wherever it wants to.

So the next time you consider whether or not you are treating an inanimate object too roughly, please keep in mind that its component particles are all legitimate Big Bang Royal Family (BBRF) member MSPs just as deservingly as you and I and all other obviously animate MSPs, including your beloved family pet(s) (if you have any), are such.

Reference

1. Folger, T. A universe built for us. *Discover*, Dec. 2008: 52–58.

Chapter Four

How Pleasure-Producing Drugs Might Be Harnessed to Enhance and Automate Learning Abilities and Processes, Work Skills, Dieting, Bodybuilding Exercise, Sleep, Etc.

Virtually all pleasure/euphoria-producing and/or addictive drugs (including alcohol, nicotine and, broadly speaking, all stimulants, sedatives, narcotics, etc.) could be used together with electromagnetic/acoustical brain stimulation (BS) to enhance and automate learning, working, dieting, exercising, sleeping, etc., by using BS circuits designed in such ways as to enhance the drug's attractive/addictive qualities and effects if and only if and when and only when the student/worker/dieter/exerciser/sleeper, etc., is (according to their own specific mental-and/or-physical-activity-indicative and characteristic brainwaves) engaging in the behavior/experience that she or he desires to engage in and improve (ref. 1).

The BS circuit should significantly increase the drug-induced pleasure when he or she is engaging in the desired behavior/experience and significantly decrease this pleasure when he or she is not engaging in it. The BS circuit should also be designed in such ways as to completely inhibit those areas of the brain that are involved in / subserve cravings for the drug(s) in question, so as to preclude actual addiction.

And if the frequency of the use of the drug(s), even in association with the constructive activities listed above, entails some adverse health effects, then designer drugs that have the same or similar pleasure-producing effects (without the deleterious health effects) should be used in place of / instead of the original drug(s) whose pleasure properties are intended to be mimicked.

Consequently, all, most or many of the drugs that seem to have been virtually nothing but troublesome could be harnessed in such ways as to be nothing but helpful to humankind.

Reference

1. Butler, S. R., Giaquinto, S. Technical note: stimulation triggered automatically by electrophysiological events. Med & Biol Engng, changed to Med Biol Comput, 1969; 7: 329–331.

Chapter Five
What Will Become of Planet Earth After It Is De-Hellified and Heavenized Around the Year 2100?

Planet Earth will survive as a museum and reunion meeting place for all animate and inanimate Earth-Hell graduates. And an infinite number of non-graduate MSPs will be more than happy to take turns serving in ongoing animate (i.e., fully conscious) and inanimate (i.e., subconscious) roles within the hallowed museum of planet Earth.

What is meant by "non-graduate MSPs" is particles who never resided on or in Earth-Hell and/or its associated Purgatory. Hence, these will be visitor MSPs who are interested in Earth for its historical value to the infinite universe, which is infinite Heaven.

Chapter Six
Some Promising Developments That Point Us toward the Time Period of 2009 to 2050

Reproducible or potentially reproducible out-of-body experiences (OBEs) that could lay the groundwork for global MSP circulation are already being produced by weak electromagnetic fields directed at the brain's temporal lobes, especially when coupled with administration of the anesthetic, ketamine (ref. 1) stimulation of the brain's temporal-parietal junction (a part of the brain that is important in body orientation in space [ref. 2]) and stimulation of the brain's gyrus angularis in the parietal lobe (ref. 3).

Also already possible and feasible is noninvasive brain stimulation that could be used to excite (or inhibit) the brain's pleasure centers by means of ultrasound transmitted through air, without the need for coupling gel or water as conductive media (refs. 4–7) and/or by means of multiple electromagnetic coils positioned strategically around the outside of the head (ref. 8). Another possible way to stimulate anywhere in the brain would be via focused, intersecting light (i.e., photon) beams (ref. 9).

And it is now well established that the brain's beta waves (frequency range: 12 to 30 Hertz or Hz) and gamma waves (above 30 Hz), which indicate the occurrence of focused attention (i.e., concentration) and problem-solving mental activity (refs. 10, 11) could potentially be used to control an electronic device such as a computerized (i.e., computerized, brain chip-containing) noninvasive brain stimulator, whereby a person's IQ could conceivably be augmented a millionfold (ref. 12), learn new languages instantly (ref. 13) and take great pleasure (and thereby become enormously motivated, knowledgeable and technically competent) in terms of both learning abilities and diversified work skills.

"The notion that people love doing things because they're good at them is back to front" (backwards). Rather: "They're good at them because they love doing them" (ref. 14). And "it seems we all have the potential for genius" (ref. 14).

Moreover, "experts" on the subject of euthanasia are at least talking about "Voluntary Euthanasia … (becoming) … legalized … whereby … medically-assisted … (suicide) might provide … a … (legitimate) means of escape from … (insurmountable-life-circumstantial) … difficulties" (ref. 15).

References

1. Kotler S. Extreme states: out-of-body experiences? *Discover*, July 2005; <u>26</u> (7): 60–67.
2. Bosveld J. Soul search. *Discover*, June 2007; special issue: 46–50.
3. Hoppe C. Controlling epilepsy. *Scientific American Mind*, June/July 2006: 62–67.
4. Basu P. Sterilized by sound. *Discover*, May 2003: 13.
5. Fink M. Time-reversed acoustics. *Scientific American*, Nov. 1999: 91–97.
6. Davies J. Prophetic patent. *New Scientist*, April 23, 2005; <u>186</u> (2496): 31.
7. Hogan J., Fox B. Sony patent takes first step to real-life matrix. *New Scientist*, April 9, 2005: 10.
8. George M. Stimulating the brain. *Scientific American*, Sept. 2003; <u>289</u> (3): 66–73.
9. Griffith A. Chipping in: brain chip for memory repair closes in on live tests. *Scientific American*, Feb. 2007; <u>296</u> (2): 18–20.
10. Sellers B., Ortiz, D. Neural interfacing and its implications on psychology. Wikipedia, the free encyclopedia, Electroencephalography, Feb. 25, 2006: sec. 2. Retrieved Feb. 17, 2006 from http://en.wikipedia.org/wiki/Electroencephalography.
11. Fox D. Brainwave boogie-woogie. *New Scientist*, Dec. 2005; <u>188</u> (2531/2532): 50–51.
12. Kurzweil R. The coming merging of mind and machine. *Scientific American Reports*, May 6, 2008; <u>18</u> (1): 20–25.
13. Horgan J. The myth of mind control. *Discover*, Oct. 2004; <u>25</u> (10): 40–46.
14. Webb J. The sky's the limit. *New Scientist*, Sept. 16–22, 2006; <u>191</u> (2569): 3.
15. Young E. Choosing to Die: Electric Death and Multiculturalism by C. G. Prado, NY, NY, Cambridge Univ. Press, 2008 & Medically Assisted Death by Robert Young, NY, NY, Cambridge Univ. Press, 2007. JAMA, Oct. 8, 2008, book and media reviews; <u>300</u> (14): 1703–1704.

Chapter Seven
Foolproof Treatment for Pedophilia and Prevention of Sexually Transmitted Diseases, Unwanted Pregnancies, Abortions, Erectile-Dysfunction-related Divorce, Rape, Etc.

Turning attention to the always controversial subjects of sex and romantic love, Drs. Robert G. Heath and Charles E. Moan of the Departments of Psychiatry and Neurology at Tulane University, New Orleans, Louisiana, discovered in 1972 that they could induce heterosexual behavior in a 19-year-old (previously exclusively) homosexual man via stimulation of the septal region of the brain in such ways that a "lady of the night" allegedly/anecdotally said that he was "the best (she'd) ever had" (ref. 1). And he stimulated his brain some "1500 times, experienced an almost overwhelming euphoria and elation and had to be disconnected, despite his vigorous protests" (refs. 2–4).

Based on their work with this young man as well as work done with other men (and women), they deduced that "people (apparently irrespective of gender and 'orientation') actually preferred" (solitary or coupled) "direct stimulation of the pleasure centers to having" (at least casual loveless) "sex with another person" (or persons?) (ref. 4). Therefore, they reasoned that electrical brain stimulation (EBS) could be used to "cure" either homosexuality or heterosexuality (ref. 4).

From these observations and inferences, we can deduce that brain stimulation (BS) might also be used to cure pedophilia and enhance the efficacy of Viagra, etc., vis-à-vis erectile dysfunction (ED) in married men or otherwise coupled men, thereby decreasing the frequency and rationale for divorce or separation. Moreover, if people actually prefer solitary, sexually pleasurable BS to casual, loveless sex, then it could be used to prevent every sexually transmissible disease from body lice to AIDS, as well as unwanted pregnancies and abortions.

Furthermore, since BS could be used either to excite or to inhibit the brain's sexual (or any other kind of) pleasure centers and/or pathways, it could be used not only to (a) enhance sexual gratification in anyone who wants it, but also to (b) eliminate sexual compulsions in anyone who doesn't want to be bothered by, waste time with and/or make foolish life-circumstantial decisions that are driven by tyrannical sexual impulses. So, it might also be useful in preventing the crime of rape.

A temporarily final thought along these lines is that BS might be used to simultaneously induce (a) an out-of-body experience (OBE) and (b) one or more intense orgasms, so as to bring about

out-of-body-sex (OB-SEX or OBS) that would leave the person in question's body completely unsoiled, un-sweated-up and uninvolved/unstressed cardiovascularly, musculoskeletally and in all other readily observable ways.

References

1. Dowd D. Instructor, Tulane Univ. School of Medicine; personal communication, 1992.
2. Moan C., Heath R. Septal stimulation for the initiation of heterosexual behavior in a homosexual male. J. Behav. Ther. & Exp. Psychiat. Vol. 3, pp. 23–30. Pergamon Press, 1972. Printed in Great Britain.
3. Baumeister A. Serendipity and the cerebral localization of pleasure. Journal of the History of the Neurosciences, Volume 15, Number 2, June 2006, pp. 92–98 (7).
4. Dr. Robert Heath (1915–1999). http://www.wireheading.com/robert-heath.html. Page 2 of 3.

Chapter Eight: The World Is Unfair
The world is an unfair place, but its unfairness will be partly corrected for right here on Earth and fully corrected for in the afterlife.

The world seems to tell us that a relatively small minority of us are worthy of being fabulously famous and wildly wealthy while the vast majority of us are *not*. Why is this? It doesn't make sense.

Is it because the elite rich and famous are intrinsically superior to the rest of us? Are the privileged few simply more intelligent and capable than the "ordinary" majority of the world's population? Or were and are the super-rich and super-famous simply lucky people who have been and are "in the right place at the right time"?

It seems to me that the intrinsic superiority argument is a load of nonsense. And the circumstantial "right-place, right-time" explanation is the correct one. Furthermore, I admit that a large part of *my* problem is simply that I'm *jealous* of the super-rich and super-famous.

But is it all a matter of purely random or "dumb luck"? Or is the hostile universe at large *actively* preventing *all* of us humans from being super-rich and super-famous (and super-intelligent, etc., too)? Is the hostile universe at large actively and deliberately sowing the seeds of apparent inequality because they want to sow the seeds of jealousy, strife and low self-esteem among the "ordinary" majority of people? I believe the malicious universe explanation much more than the random "dumb luck" explanation.

Because if there weren't powerful forces in the universe at large who actively promote and protect the super-rich and super-famous from the rest of us, we relatively underprivileged "ordinary" folks would rise up against the overprivileged minority and literally destroy them.

But *why* would the universe be so malicious as to favor a small minority of people at the expense of the majority of us? I think the reason why *they* (the infinite number of angry extraterrestrial mind particles) do this is because they figure:

"Look, if we extraterrestrials can tolerate our jealousy of the whole lot of *all* of your earthlings' being center-stage to the universe, then the majority of you earthlings should 'get a taste of your own medicine' by your being forced to tolerate *your* jealousy of the overprivileged minority of your earthlings. 'Misery loves company.' So, if we extraterrestrial MPs have to live with *our* jealousy of *all* earthlings, then we want to at least be entertained by seeing the majority of you earthlings feeling depressed and jealous in relation to some small minority of *your* population!"

After we die, our MPs leave the universe's center-stage and go to a peripheral-to-planet-Earth, scooped-out (by Earth) spherical volume of space that surrounds our planet. This volume might more or less comprise the 15 or so billion light-years of the visible universe that encircle our world (ref. 1).

Within this peripheral-to-Earth volumetric region, all of our deceased loved ones and other graduates of planet Hell (Earth) may be working on behalf of *all* of us (underprivileged *and* overprivileged) earthlings who are still trapped here in Hell. They're trying to negotiate with the hateful infinite hordes of extraterrestrial MPs who want the majority of earthlings to be as unhappy and dissatisfied with our terrestrial status as they are unhappy and dissatisfied with *their* extraterrestrial (lack of center-stage) status.

Once the world has been fully de-Hellified and Heavenized, within the next 50 to 100+ years, we currently "living" earthlings will join up with our "deceased" friends and relatives who are among our Big Bang Family (BBF) members.

What exactly is a BBF? It is the entire group of mind-soul-particles (MSPs) who were chosen (by God?) at the time of the most recent Big Bang Explosion to populate the universe's central-reference-point (CRP) Hell. So our Big Bang Family consists in the population of all past, present and future graduates of planet Earth-Hell.

Once *all* members of our BBF have "lived," "died" and left planet Earth, the universe at large will no longer be hotly jealous of us. This is because, by that time, we will have departed from the universe's center-stage focal point (planet Earth). We currently "living," bonded-to-Earth MPs will be thrilled to be reunited with our "deceased" fellow BBF members.

This reunion will take place within the Earth-center-scooped-out spherical volume of the universe that encircles and surrounds Earth within the 15 (plus or minus) billion light-year region of the universe that is observable and which we might as well call "Purgatory."

So, despite what astrophysicists observe in our deliberately misleading, Earth-bonded context, what the universe really amounts to is a central-reference-point (Earth) which is Hell, surrounded by a *finite* spherical volume which is Purgatory, surrounded by an *infinite*, boundless realm which is Heaven.

Purgatory is where all of the graduates of Earth-Hell, i.e., "deceased" earthling-BBF-members dwell and *hover* (actively, protectively and helpfully) *over* the rest of us BBF-members who are still trapped here in CRP-Earth-Hell. They, the Purgatites (their MSPs) probably comprise the "invisible," "dark matter and energy" of the universe, which astrophysicists are fond of thinking and talking about (ref. 2). The Heavenites are the angry ones who hate all earthlings because they are so jealous of our CRP-status that they don't even want us to be aware of it.

Purgatory lies outside of and surrounds the center-stage *focal point* of the universe (i.e., Earth = Hell). So the infinite number of extraterrestrial Heavenites *do not like* but *do not hate* the finite number of "deceased" Earth-graduates nearly as much as they hate us still-"alive" earthlings.

And as soon as we still-"alive," Hell-central-focal-point-within-the-CRP (Hell and Purgatory)-dwelling earthlings are all "dead," i.e., our MPs have *left* Hell and returned to the loving "arms" of our "deceased" fellow BBF members, then we'll *all* (all current BBF members) be free to leave both Hell and Purgatory altogether. Thence, we'll all return to Heaven, where we will enjoy the permanent status of former-BBF-Hell-Purgatory graduate celebrities. No wonder that those Heavenites who have never been to Hell-Purgatory, at least not as residents thereof, and have never (yet, anyway) been BBF members, are jealous, hateful and malevolent toward us. And they are the source and reason for all of our pain and suffering. How they afflict us is beyond the scope of this manuscript, except to say that they do it both by withholding things that would help us and by inflicting upon us things that hinder and harm us.

And once we current BBF members are all "dead," we will leave Hell and Purgatory as one harmonious, jealousy-free BBF. We will be one "big happy family" because all of the inequities of fame, money, "intelligence," etc., will be corrected for by a gigantic device that might be called a "cyclocosmotron" whereby each *under*privileged, "ordinary" (i.e., non-worldly-famous, non-super-rich) Earth-graduate will be allowed to travel around and among the whole BBF in much the same way as the Queen of England travels around and among her "subjects," i.e., as the center of attention.

So, via the cyclocosmotron, which is a gigantic mind-soul-particle (MSP) accelerator and trajectory-director, all members of our BBF will end up getting the *same* amount of cumulative attention (i.e., fame). Needless to say, the Queen of England, Albert Einstein and other *over*privileged-individual, Earth-graduated MPs will either get *no* turn to ride the cyclocosmotron or only a very brief turn. As my priest at the Roman Catholic church is fond of saying: "And the last shall be first, and the first, last." (Matthew 20:16, The Living Bible: Tyndale House Publishers, 1987: Matthew, 18.)

Then, there will be no more jealousy among us planet-Earth graduated BBF members. Consequently, we will all *love* each other. Once the maldistribution of attention-fame, etc., has been corrected for, we Earth graduates will all travel around the infinite universe as permanent major celebrities. We'll all be famous throughout infinite Heaven at large because, unlike the infinite majority of Heavenites, we will have the distinction of having been to Hell and Purgatory and back (to Heaven). We will be able to travel through Heaven either singly or in groups of our fellow Big Bang Family members. And we *will* have family reunions at intermittent intervals *after* we graduate from our Hell and our Purgatory.

So, another reason why Heavenites, who have never been chosen (by God?) to live on a CRP-Hell planet (a new one is selected with each successive Big Bang Explosion), want to see us suffer (i.e., pay our dues to the universe to which we have the privilege and burden of being central) is that they feel not only should we have to entertain them if they're going to pay any attention to us but also we should have to *earn* our permanent, eternal, post-Hell, Heavenly celebrity status (as graduates of Hell and Purgatory).

There is much less pain and suffering in Purgatory than in Hell. The only pain and suffering in Purgatory is that connected with the burden of having to help those of us BBF members who are still "alive" (i.e., currently living on Earth-Hell) to endure and to tolerate our bondage here in Hell. The residents of Purgatory are twofold in kind:

(1) those BBF members who have not yet been chosen to live and "die" on Earth-Hell and

(2) those BBF members who have already lived and "died" on Earth-Hell.

Hence, Purgatory is populated by a combination of the yet-to-be-born and the "dead" earthlings, including all of our "deceased" loved ones and friends. These Purgatites, who are all members of the current BBF, spend most of their awake time striving to help us by begging the angrily jealous Heavenites to be merciful to us earthling-Hellites and by intervening in Heavenly ways in our Hellish affairs/problems. Purgatory is not a *wonderful* place like Heaven. In (infinite)

Heaven the only thing that is *not* Heavenly is the hateful jealousy and resultant wrongdoing/evildoing that is directed at us "still alive" earthlings.

Purgatory is a place where all of the current BBF members who have not yet been allowed to act out their parts in the play of Earth-Hell-TV *work* alongside all of the current BBF members who have already played their parts in the play on Earth-Hell-TV. So, in Purgatory the yet-unborn and the already-lived-and-"died" BBF members work together on behalf of all of us BBF members who are currently acting out our respective Earth-Hell-TV parts in the cosmic CRP play. All of the Purgatites have, as their cosmic duty, the task of trying to help us "living" BBF members by exerting forward-directed pressure and negotiating the pace of "progress" with our infinite number of envious, Heavenly enemies.

So, there are three categories of current BBF members:

(1) Yet-to-be-born members who reside in Purgatory,
(2) Currently-"living" members who live on Earth-Hell, and
(3) Already-"dead" former or ex-earthlings who reside in Purgatory.

In both Purgatory and Heaven, there is no need for money because goods and services are so plentiful that everyone in Purgatory and Heaven can have whatever they want in the way of material goods and services whenever and as often as they want them at no (zero) monetary cost. Money is simply part of the *"hell"* of Earth-Hell. In both Purgatory and Heaven, no one ever worries about loss or destruction of property because it can always be replaced as soon as or sooner than it can be lost or destroyed.

And for those unsympathetic diehards who would argue that here on Earth, money and even fame is/are proportional to learning and work-related effort, a large dose of common sense is needed. Learning and work accomplished are proportional not only to effort, but also to pleasure (i.e., the amount of "interest") taken in the big-money-fame-earning activities in question and inversely proportional to how much pain (mental and/or physical) is associated with making the effort connected with the learning or working. So,

$$LW = \frac{E \times Pl}{Pa}$$

where L = learning, W = work accomplished, E = effort put forth, Pl = pleasure or interest taken (for example, in big-money/fame-earning activities) and Pa = pain associated with the effort involved in any learning or working, as we currently know it to be. But since pain is inversely proportional to pleasure (Pa = 1 ÷ Pl), you can substitute one divided by Pl for the Pa term. Consequently,

$$LW = \frac{E \times Pl}{Pa} = LW = \frac{E \times Pl}{\frac{1}{Pl}}$$

which = $E \times (Pl)^2$.

So, LW = E (Pl)^2, or learning/work equals effort times pleasure squared, which *is* analogous to E = mc^2 for inanimate (unconscious and subconscious matter). But LW = E (Pl)^2 is only an approximation which pertains to conscious matter (i.e., matter that has had its consciousizer mechanism turned on by pleasure).

The reason why LW = E (Pl)^2 is only an approximation is because it is, strictly speaking, a *proportionality* relationship rather than an equation in the strict sense. The mathematical relationship between LW = E (Pl)^2 and E = mc^2 is beyond the scope of this manuscript but may be analyzed and discussed in a sequel (manuscript) to this one. For now, suffice it to say that for any MP, when it is in its unconscious or subconscious state its energy, E = mc^2, but when that same MP becomes conscious, then its energy level (which is proportional to its learning/work level) is proportional to its effort times its pleasure squared. So:

E MP-uncon/subcon = mc^2, but

E MP-con = LW = E (Pl)^2, where

the first "E" denotes "energy" or "energy level" and the second "E" denotes "effort."

In any case, what this incomplete equation or proportionality relationship expresses is the notion that the super-rich and super-famous people (here *within* Hell) are not super-rich and super-famous because they are putting forth some gigantic effort in their learning and work processes, but rather because they're deriving gigantic pleasure from (i.e., are enormously interested in, hence, greatly enjoying and motivated in relation to) whatever lucrative subject matter or endeavors that are bringing in the enormous money and fame/attention. And they are doing this so much so that they need to put forth only relatively minuscule effort in order to acquire huge dividends in terms of money and fame/attention. *Hence, they are being rewarded for having fun* (i.e., pleasure) and do not necessarily *deserve* to be so richly rewarded on the basis of any meager pain that they actually need to endure.

It is highly paradoxical that *all* Hellites (and Purgatites) are infinitely famous throughout the infinite universe, yet only a tiny percentage (0.01%?) of Hellites are famous (enough to be designated "celebrities") here *within* finite Hell. In contrast all Purgatites are equally and enormously famous throughout Purgatory and the rest of the infinite universe. And all Hellites are famous throughout finite Purgatory as well as throughout (infinitely large) Heaven.

So, even if you're not famous within the world, you are *infinitely* famous throughout (finite) Purgatory and infinite Heaven. So, here in Hell we are *deliberately not famous* (except if you're among the lucky 0.01±%) because the Heavenites are too jealous of us to let us know about our infinite fame until we leave the limelight of Hell and enter into the more diffuse (but still-part-of-CRP centrally-focused) light of Purgatory, where they can tolerate what they see as our smugness about being infinitely famous BBF members.

So, here in Hell our being infinitely famous is being kept a secret (a well-kept secret) from us by the jealous Heavenites. In other words, here in Hell, we occupy the "secret center" of the universe. Again, it seems an uncanny coincidence that my favorite teacher, Louise Jameyson titled one of her paintings "Secret Center," almost as if she had prior knowledge of an idea that (God? and) the Purgatites were going to let *"come to"* me *after* Jamey's "death" (graduation to Purgatory) in 1997.

22

Chapter Nine: To Hell with Heaven If We Can't Limit Hell
(and What about Space Aliens and UFOs?)

Part of how we can intuit that this world is Hell has to do with disease, pain and suffering. But another part is the observation that our pains hurt much more than our pleasures feel good. For example, two people can be friends for decades and then, if one friend hears the other saying something critical of the first one, the friendship can be permanently ended and over with. The same can happen if one friend emotionally scolds the other one on just *one* occasion, despite subsequent profuse apologies.

And when we feel comfortable, we feel fine, but when we don't feel comfortable (as when we're in pain) we can feel *desperately terrible*. Similarly, when we have enough money for our needs and reasonable wants, we feel good. But when we don't have enough money, we feel *absolutely terrible* and absolutely terrible things can happen to us. Moreover, consider how *difficult* it is (i.e., how much time and effort it takes) to make almost *any* product of human labor compared to how *easy* it is (i.e., how little time and effort it takes) to destroy almost *any* product of human labor.

If Earth were just a *neutral* place as opposed to a bad one, wouldn't our pleasures be just as intense and durable as our pains? Two of my favorite teachers pointed out (independently of each other; in 1967, Miriam Goldeen was one of my high school teachers, and in 1971, Blanchard Means was one of my college-level instructors) that "no one knows why" our pains and afflictions (here on Earth) hurt so much more than our pleasures and blessings feel good. I say it's because the planet is Hell.

What about a thousand years ago? Wasn't that *worse* or more intense Hell? Of course it was. Moreover, "life" is much more pervasively boring and tedious than it is pervasively ecstatically or even highly pleasureful. On the contrary, life here seems to be little more or better than a series of *problems*. It's rare that we have *problem-free*, pleasurable periods. Everyone knows the world is basically a cruel place. So, I say it's Hell!

And the only way our Hell could be stable (i.e., pleasurable) enough to exist at all or be as predictably good as it is *is* if it is being stabilized from the *outside* by pain-controlling and pleasureful phenomena trickling in to us from an enormously (or infinitely) more powerful and more stable realm or realms such as Purgatory (where we have friends) and Heaven where (except for God?) we have an infinite number of souls (MSPs or MPs, *who are intrinsically no worse than we are*), but who because of their cosmic envy of our cosmic centrality are, at best, lukewarm, little-caring friends and, at worst, *monstrous enemies*.

So, if Heaven exists, why doesn't it just come here and announce itself to us? The possible answer: Because infinitely-large Heaven is such a wonderfully painless, suffering-free and pleasureful place that if we *knew* what we were missing, *we would all rush around* frenetically *to commit suicide* as soon as we could, in almost *any way we could*!

And if all of us here in Hell were to commit suicide, then the universe would *lose* its Hell-television, entertaining central-reference-point and then the universe would become too chaotic both chronologically and spatially for the universe (including Heaven) to exist at all! Without this Hell, there could be no Heaven.

We're doing the universe a big favor by tolerating this life at all. But I say "enough is enough" and we should be able to *limit* how long we have to be here! To hell with Heaven and the universe itself if we have to put up with all of this boredom, pain and suffering for 100+ years. I say 60 or 65 years would be more than enough for *me*. And I'm 57 now. Oh well, Hell goes on for now.

If planet Earth weren't unique in terms of being dominated by pain and suffering (as opposed to pleasure and joyfulness) it would merely be one of an infinite number of happy planets populated by basically happy MSPs. And it would therefore have no special attention-grabbing characteristics and therefore could *not* serve as the cosmic universal central-reference-point (CRP). After all, how would we be entertaining to the rest of the infinite universe if we earthlings were simply having a wonderful time here? What fun would we be for them as members of our infinite audience?

If we were all having a wonderful time, we would be as ordinary as the infinite number of happy planets that are out there in infinite Heaven. And no one out there would pay any special attention to us. Hence, we need to suffer here on Earth in order to be attention-worthy in the "eyes" of the infinitely large and happy universe. However, we should be able to determine *how much* pain, suffering and unhappiness we're going to stand for.

Moreover, since the infinite number of basically happy MSPs who populate infinite Heaven are basically hostile to us earthlings because of our center-stage status, it is valid for us to think of them as an infinite number of *"space aliens"* who are our enemies. And the Existentialists' notion that the universe is basically hostile to us (ref. 1) is a valid one. Plus, the popular notion that unidentified flying objects *(UFOs)* are often genuine as such and are manned/womanned by hostile space aliens may very well be a valid one.

However, the Existentialists could not be further from the truth in their contention that our earthly lives are devoid of intrinsic and ultimate purpose. As *this* book contends, our purpose here on Earth/Hell is to provide a CRP for the infinite universe without which infinite Heaven could not exist as an orderly, harmonious and essentially stable place, hence, without which the universe (everything that exists) might be too unstable to exist at all.

Reference

1. Crystal, D. *The Cambridge Encyclopedia*, Cambridge University Press, 1994: 402.

Chapter Ten: The Universe Is Infinitely Large and Old

The universe is infinitely large, old and contains an infinite number of immortal particles (all of which can be either [1] conscious-and-animate or [2] unconscious/subconscious-and-inanimate) that flicker in and out of existence.

It stands to reason that the universe is infinitely large. This is because if you imagine some outer boundary to the universe and then ask yourself what might lie outside that boundary, the obvious answer is "more space."

It also stands to reason that the universe is infinitely old. This is because if you imagine some initial point in time when the universe first came into existence and then ask yourself what existed before the universe came into being, the obvious answer is *infinite* empty(?) space and *time*.

We know that matter exists in our world, that is, in the space that we humans (can) occupy. So why couldn't or wouldn't matter exist everywhere else, too? Hence, let us hypothesize that the universe is infinitely large and old and contains an infinite amount of matter, which consists in an infinite number of particles which, like our own minds (i.e., MSPs or MPs) can exist as either (a) conscious, hence, animate matter or (b) unconscious or subconscious, hence, inanimate matter, such as when we're asleep or under general anesthesia.

All of science tells us that matter is made up of particles that are too small to be seen, at least with our naked human eyes. And a relatively new theory called "superstring" or "string" theory tells us that *everything* is made up of string-like particles that are 10^{-33} meters long (ref. 3) and have zero thickness (ref. 4). This is impossible because anything with *zero* thickness also has zero length.

So, why should we bother with string theory at all if it tells us even one thing that is logically impossible? Because this theory seems to *unify* (render consistent with each other) all of the mathematical equations which underlie the deepest field of science, that is, physics.

Okay, so let's modify string theory by allowing as plausible that these hypothetical, ultimate, elemental building blocks of nature may have a length of 10^{-33} meters. However, in order to exist at all, their thickness cannot be zero and must, instead, be some very small but *non*-zero number or value.

If *everything* is made up of superstring particles, then you and I, as *individual*, hence unitary entities of *some* kind, must each *be* a superstring particle that is trapped inside a human body (made up of many superstring particles) for some reason. So, you *are* a single, most-of-the-time *conscious* superstring, but your body (including your brain) is made up of a huge number of unconscious or subconscious superstring particles.

So, you're a superstring particle who is imprisoned within a problem-prone body. I am too, and so is every other individual, unitary, conscious (human *or nonhuman*) being.

Now let's suppose that you and I and every other individual superstring (particle) can flicker in and out of existence, similarly as we can flicker in and out of a state of awareness versus subconsciousness/unconsciousness whenever we go back and forth between being awake and being asleep. Let's theorize that each superstring-individual or individual-superstring can do all of the following:

    (1)   exist and be conscious, hence animate;

    (2)   exist and be subconscious or unconscious, hence inanimate;

    (3)   disappear entirely, without violating any physical law of conservation of mass, matter or energy, because by disappearing you're simply converting *actual* mass, matter and energy into *potential* mass, matter and energy; and

    (4)   reappear at any time it *wants* to.

Looking through time past, you never existed for the first time and you'll never exist for the last time. Hence, you're eternally (although intermittently) immortal in both past and future directions of time.

Just because you and I *can* have a length of 10^{-33} meters and *almost* zero thickness, doesn't mean we're *limited* to that size and shape. Let's hypothesize that you and I (and every other individual, unitary particle) can vary our size anywhere between zero (*nonexistence*) and *infinity* and can vary our shape in any and all ways in which we want to. So, all particles of matter have an infinitely variable size and shape.

Chapter Eleven: You're Not Your Brain
Your brain and you (i.e., your mind) are two different things.

Think about it: Your brain is made up of zillions of particles, yet you know yourself to be just one individual entity, hence *one* something. People even refer to you as an "individual." And if everything in our universe consists in one or more particle(s), then the one, individual something you are must be one, individual particle. So, your brain is made up of zillions of molecules, atoms and subatomic particles, all of which consist in *elemental building-block particles* (EBBPs), which we are hypothesizing to be superstring particles.

You are your mind, which is just *one* thing, that is, one superstring particle that has its consciousizer mechanism turned on, except for when you're neither awake nor engaging in dream-pervaded sleep. Let's hypothesize that your unitary mind is the same thing as your unitary "soul." Your mind may be referred to as either a mind particle (MP), a soul particle (S) or a mind-soul-particle (MSP). And MP, SP and MSP are three synonyms for each other.

Your MSP is one most-of-the-time consciousized particle which is different from and intrinsically separate from your brain, which is an aggregate or conglomerate of subconsciousized or unconsciousized zillions of individual superstring particles.

While out-of-body experiences are still relatively uncommon and rare, there are two commonplace phenomena that suggest that the mind and body (including the brain) are two different things.

1. When you're asleep and dreaming, you're well aware of your mind's activity, but virtually unaware of any part of your body's (including your brain's) position, bodily sensations, etc., and
2. If you've ever experienced <u>orthostatic hypotension</u> (i.e., "lightheadedness" and dizziness) when arising suddenly from a sitting or squatting position, you might realize that this feels somewhat as though your mind is separating from your body (including your brain).

Furthermore, the idea that the mind and brain are two different things entails <u>an uncanny coincidence</u> between (a) the realization that your mind is just one thing, whereas your brain is made up of many component elements and (b) the religious notion that the mind and the soul are one and the same experiencing entity or <u>experiencer</u>. Actually, it is contended herein that this coincidence is more than just a chance-based coincidence and reflects the deep-seated reality that the mind (a mundane concept) and the soul (particle) (a religious concept) are one and the same aspects of <u>a single reality that has</u> <u>both scientific and religious aspects</u>.

Chapter Twelve: All Minds Are Created Equal
Why all minds are created equal, yet (equally) unique, even though all brains are not equal or equally unique.

If all MPs *weren't* created *equal* in overall quality (the sum total value of learning ability, pleasure-experiencing ability, kindness, courage and all other characteristics that everyone values), there would be so much jealousy, rivalry and hostility in the universe at large, that it would be too unstable (pleasureless) to exist at all. And all MPs would cease to exist out of a desire to avoid constant hatred, conflict, pain and suffering, hence instability.

Thus, all MPs are created (or spontaneously pop into existence) in a condition of equal sum total quality. In other words, one MP might have *more* learning ability, for example, than another one, but then it might also have *less* kindness or less of some other valued characteristic(s), so that the sum total of universally valued characteristics is the same or equal between any two MPs and among all MPs in the universe. Hence, no one is any better than or superior to anyone else in the universe at large, despite the ugly and painful, hence, destabilizing situation that we have here on Earth-Hell, that some people seem to be so much more favored in terms of fame, money, "intelligence" and all other worldly-valued characteristics than others.

The fact that this criterion of sum total quality-equality is not met here on Earth (for example, some people have better brains and are otherwise more fortunate than others), doesn't subtract from the argument that real or apparent inequality leads to hatred and conflict. On the contrary, the consequences of this inequality support and strengthen the argument. Thus, what's true here on Earth (i.e., inter-individual inequality) cannot possibly be true in the universe at large, because if it were true, the universe would be too (painful, hence) unstable to exist at all and all you would have would be an infinite amount of empty space and time. How it is that the world can be stable enough to exist despite the hostility that results from apparent inter-individual inequality is explained below.

However, all MPs are created (or spontaneously pop into existence) as *equally unusual* and *unique* entities. This is because if we were not each intrinsically *equally* extraordinary and unique, hence, *equally special*, we would all be so jealous, hostile and depressed, hence unstable, that we would cease to exist.

The only reason why unstable Earth can seem as stable, hence predictable as it (barely and hardly) does, despite glaringly obvious inter-individual inequality, is because Earth is constantly being stabilized by imperceptible (to us, anyway) *reassuring* input from the benevolent, egalitarian, Purgatistic realm of all yet-unborn and all already-"dead" earthlings which might encircle much, most or all of the full volume of visible/observable space surrounding planet Earth (ref. 1, p. 94).

Chapter Thirteen: Almost Infinite Number of Spatial Dimensions

There are an (almost) infinite number of dimensions of space (and time).

Despite superstring theory's contention that there are only a finite number of spatial dimensions (9, 10, 11, 24, 25, 26, etc. ?, ref. 5, p. 94), there must actually be an (almost) infinite number of dimensions of space. This is because you (apart from the body in which you're temporarily trapped) can make yourself (almost) as thin as you want to be. This is because you're a size-variable superstring or MSP. The reason for the "almost" qualification is that you can never simultaneously exist and have absolute *zero* thickness or *infinite* thinness.

So, an (almost) infinite number of MSPs could pass through an (almost) infinitesimally small spherical volume of space without ever touching or even being aware of each other's existence. Hence, there are an almost infinite number of mutually exclusive dimensions of space that are inherent in each and every spherical point (within any [nearly] infinitesimally small spherical volume of) space.

The number of dimensions of time is also (almost) infinite but is a trickier subject to discuss, because it involves both time expansion/dilation (i.e., FORWARD RELATIVITY) and time contraction/shrinkage (i.e., REVERSE RELATIVITY) (ref. 6, p. 94), which will be postponed until and unless some sequel to this manuscript comes into being.

Chapter Fourteen: Necessary Central-Reference-Point

Why the universe must have an arbitrarily designated central-reference-point.

Although the universe is infinitely large and therefore cannot have a true geometric center, in order for any two or more MPs to make any plans to do *anything* together, all MPs must agree on some arbitrarily designated *central-reference-point* (CRP) with spatial coordinates:

$$x = 0, y = 0 \text{ and } z = 0.$$

Without a CRP, no two or more MPs could ever cooperate or work together on anything because they could not possibly propose any joint venture(s) such as: "Let's meet at time, t = a, (where 'a' is the amount of time elapsed since the most recent Big-Bang-Explosion-mediated choice of CRP-planet) and at spatial point/location, x = a, y = b and c= z."

Chapter Fifteen: Why Central-Reference-Point Must Be Hellish

Why CRP must be Hellish.

Whether we admit it or not, we all want to be famous. Hence, if the *Hellishness* within the CRP weren't there (here? on Earth) to act *as a deterrent*, then there could be no CRP because an *infinite* number of MPs would fearlessly, covetously and simultaneously rush toward, storm and stampede the CRP, while shouting something to the following effect: "Get out of my way, I want it, *me first*, let it be mine, *now*."

And if an infinite number of MPs were all storming the CRP simultaneously, the universe would be nothing but one infinitely voluminous, chaotic stomping ground. And the disorder, instability, pain and suffering would be so great that there could be no stability anywhere or at any time, hence there could be no existence at all. And all there could ever be, would be infinite empty space and time.

Hell is a necessary evil not only for the existence of Heaven, but also for the existence of the infinite universe itself. So, if our world didn't exist as Hell, the infinite Heavenly universe could not exist at all.

Moreover, the reason why we're burdened with our problem-prone bodies is *not* just for some *useless hell* of it, but actually for the *useful hell* of it, so that the *body-burden-free*, stability-filled universe at large, apart from planet Hell-Earth, can exist as the orderly, wonderful and Heavenly place that it is, except for the burden of jealousy (hence, an *element* of mental pain) that is directed CRP-earthward, which subsides and vanishes altogether at the end of each consecutive Big Bang cycle.

Chapter Sixteen: Why Earth Is Probably the CRP

Why planet Earth is probably the current central-reference-point (and why all matter is condensed pleasure).

If the infinite universe is made up of one kind of particle, whether we call it a "superstring" or an "MP" or anything else, then it is made up of *one* kind of matter. And if it is made up of *one* kind of matter, then, since *matter is* just *condensed energy* (ref. 7, p. 94), therefore the universe must be made up of *only one kind of energy*.

We all know that our pleasure and happiness energize and stabilize us. Thus, if there is only one kind of (stabilizing) energy in the universe, then it *must* be pleasure. Conversely, we all know that our pain, unhappiness and suffering drain our energy and destabilize us. Thus, if there is only one kind of (destabilizing) energy in the universe, it *must* be pain/suffering.

So, stability equals pleasure and instability equals pain. And the *pleasure-pain continuum* (PPC) is the *only real* and true fundamental force/counterforce or (attractive/repulsive) *force* in the universe. So, every*one* (every MP or consciousized superstring) and every*thing* (every subconscious or unconscious superstring) exists solely for the sake of its *own* (and to a lesser degree, in a vicarious, "what-goes-around, comes-around" way, its loved ones' or potential loved ones', which is to say, *everyone* else's *pleasure*.

And if there were no pleasure, nothing would exist. Without pleasure, there would be nothing but empty, infinite space and eventless, infinite time. And this is true despite the fact that *here in Hell, pleasure* (such as that associated with the "buzz" that one can get by drinking alcoholic beverages, the pleasure one can get from compulsive sexual activities, etc.) does *not* necessarily equate with *stability* of any kind.

This is because here in Hell, we are all at cross-purposes with everyone else. This Hellish reality of each person' pleasure tending to oppose everyone else's pleasure is just part of how we can rest assured that this world *is* Hell. For example, if you have the pleasure of getting drunk, then you may have and cause the pain of killing other people in a car crash. Or if you have the pleasure of casual sex, the person who *supposedly* (?) "loves" you, may suddenly stop loving you, due to their own selfish *jealousy* (which has *nothing* to do with true, selfless love).

And if there's any lingering doubt about whether or not each of us humans is basically at cross-purposes with all other humans, just consider the nature of the essence of rivalry, competitiveness, the *"me first"* and "every man (person) for himself (themself)" mentalities. In other words, we all tend to be selfish and our self-centered, self-interest ("original sin"?) tends to be at odds with everyone else's self-interest.

All of this "cynical" realism is not necessarily as valid in either Purgatory or in the universe at large (Heaven), where the emphasis is more on everyone being equal to everyone else and on *shared* (rather than *individually experienced*) *pleasures*.

So, apart from what goes on here in Hell, the answer to the question, "Why is there something rather than nothing?" (ref. 8, p. 94) is (or might probably be) that *something* and *everything* (every individual particle) pops into existence because it is (primarily) motivated by the pursuit of (its own) pleasure (and happiness, which is just a specialized or generalized form of pleasure, depending upon how you look at it).

Hence, the following expression sums up why anything and everything (as opposed to nothing) exists:

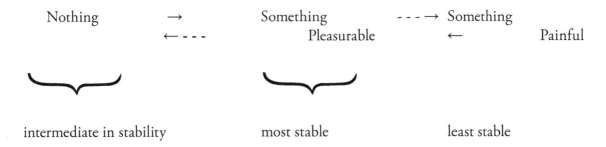

| Nothing | $\rightarrow$ | Something | $--\rightarrow$ | Something |
| | $\leftarrow--$ | Pleasurable | $\leftarrow$ | Painful |

intermediate in stability   most stable   least stable

So, *most* of infinite space is full of / populated by superstring-?/elemental-building-block particles (EBBPs)/MPs (*most* stable), an *intermediate* volume of space is full of nothingness (*intermediately* stable, and the *smallest* or *least* volume of space (probably just CRP-Earth-Hell and its surrounding environment) is relatively fraught with pain and suffering (*least* stable; most unstable).

And all four so-called "fundamental" forces of nature:

(1) the strong nuclear force,
(2) the weak nuclear force,
(3) electromagnetism and
(4) gravity must all be different modes of pleasure and stabilization. So, atoms and molecules hold together (exist as such) purely for *the* (1 through 4 or more different modes of) *fun* of doing so! Hence, the subatomic *EBBPs* within atoms and molecules *are* actually minuscule *unconscious and/or subconscious minds*, which hold and stick together only because they enjoy doing so. And the reason why subatomic particles can become furiously and explosively destructive (e.g., atomic and hydrogen bombs) is because they are intensely angry about having their pleasure suddenly interfered with and abruptly (although only temporarily) terminated!

Notice that "pleasure" and "stability" are synonyms, as are "pain" and "instability," both in relation to the mind or "psyche" as well as in relation to the fields of biology, chemistry, physics and engineering (high-technology). Moreover, IF ALL MATTER IS CONDENSED ENERGY (ref. 7, p. 94), *then, since pleasure is the only constructive energy in the universe,* therefore, ALL MATTER MUST BE CONDENSED PLEASURE.

As much as we all know that *our* pleasure energizes and stabilizes *us* personally, we also know that the world around us is *at best* neutral (if not downright *hostile!*) in its attitude and behavior toward our individual pleasure.

If a world that is hostile to your pleasure *isn't* a Hellish place, then I can't imagine what a Hellish place might be!

So, because the infinite universe operates on the pleasure principle, whereas planet Earth seems to work on an *anti-pleasure* (i.e., pleasure is problematic and exceedingly difficult and slippery to gain access to) *principle*, there's good reason to think that planet Earth (planet Hell?) is actually the singularly afflicted (bedeviled?) CRP of the infinite universe. Who selects the CRP planet? The duly-elected president of the universe? God? A systematic analysis of this question and its possible ramifications and answers is beyond the scope of this manuscript but might be within the scope of some future manuscript.

The notion that Earth might be the arbitrarily selected "center" of the universe seems to wildly and bizarrely contradict and "fly in the face" of virtually all astrophysical observations. The latter all seem to tell us that planet Earth is just a relatively ordinary, "average" planet in an ordinary solar system, in an ordinary galaxy, which is just one of trillions or an infinite number of galaxies.

I say it is because the infinite universe's infinite number of EBBPs are *just jealous* of our cosmic centrality (CC) that they arrange the physical universe's observable mega structures (planets, star-suns, black holes, galaxies, etc.) in such ways as to deflate our potential CC-based enormous egos by misleading us into *thinking* we're just average, ordinary and the universe at large doesn't give a damn about us or necessarily even know that we exist!

A possible exception to this contradiction, paradox and *perverse behavior* (i.e., *passive-aggressive behavior* on the part of the universe's *non*-BBF-EBBPs) might be the following observation/question (which does not seem to be adequately answered in the context from which it was taken [ref. 9, p. 94]): "If distant galaxies [are all]/all are moving away from us, doesn't that mean we're at the universe's center?"

As suggested above but is worth restating in other words, perhaps the reason why Earth's cosmic centrality (CC) is being hidden and kept a secret from us is that the infinite number of non-CRP, non-Earth-bonded MPs are so bitterly (vindictively, viciously, etc.?) jealous of our probable/possible CC that they do everything they can do, everything that is possible to arrange all of the various "Heavenly" bodies in such ways as to deprive us of the awareness, enjoyment and satisfaction of knowing/being aware of our potentially gratifying CC status. Plus, *we would cease to be entertaining and attention-grabbing* if we were complacent and relatively self-satisfied here on Earth.

Chapter Seventeen: Big Bang Theory Valid
Is the Big Bang theory valid?

Yes, it probably is valid, but it's probably cyclical in nature. The reasons why the universe might require a Hellish central-reference-point (CRP) might include the following:

(1) the physical need for a spatial and chronological reference point to serve as the analytic geometric centerpoint or "origin" of the boundless universe as explained above,

(2) the psychological need on the part of *each* EBBP/MSP to know that, *sooner or later, they* (he, she or, more probably *it* or *they*, since gender would seem to be a contrivance of Earth-Hell), on the basis of a one in infinity (1:∞) chance, might get selected (by God?) to be a Big Bang Family member (BBF) and thereby eventually acquire the infinite, permanent and eternal celebrity status of "someday" becoming a graduate of a Big-Bang-associated-CRP-Hell and

(3) because EBBPs/MSPs in the Heavenly universe at large get bored, irritated, weary and tired (even to the point of literally dying and becoming nonexistent) of every*thing*/ every*one* being so cloyingly, *sweetly pleasant* and *polite* all of the time, consequently, they actually *want* the *refreshment* as well as the entertainment entailed in *sour, salty, bitter pain* and *suffering*. If this is true, then William Shakespeare may have been somewhat of a physicist as well as a playwright when he wrote in his 1599 play, *As You Like It*, that "all the world is but a stage" whereupon all of the women and men get their/our turn to fret, worry, suffer, etc. (refs. 10 and 11, p. 94). Our pain and suffering might serve as *their* (the non-current-BBF-member EBBP/MSPs') comedy relief. But despite their jealousy and mean-spiritedness (original sin?), they do have the *wisdom* to *confine* the adversity and affliction to the narrowly-circumscribed CRP, where they can *monitor, sometimes* have *the decency* to *limit*, and, in general, keep an "eye" on all of the suffering.

Every few billion years or so, there might be another, brand-new Big-Bang-mediated selection of a Hellish CRP-planet which starts out as deep or stone-age Hell and gradually (via "progress") returns to the Heavenly conditions of the universe at large. The gradual return to blissful conditions occurs in proportion as the cosmic envy (CE) of those EBBPs/MSPs who were *not* included in the current, most recent Big Bang Family of CRP-bonded MPs (or any *previous* BBF in the infinite history of the universe), *subsides* (wears off) over time.

And those envious EBBPs/MSPs gradually mellow and therefore actively *input* more and more science, technology, egalitarian and altruistic philosophy into the CRP until the CRP finally returns to the "kingdom" of Heavenly conditions of the universe at large (or the "Kingdom of Heaven" finally returns to the CRP), which at the present exponentially accelerating rate of progress might be expected to happen at some time around the end of the current, twenty-first century. At that time, a new Big Bang Explosion will occur elsewhere (other than the vicinity of planet Earth) in the universe when the president of the universe (or God?) says, "And the next (Hellish) center of the universe will be right here" as S(He) points to some region of the universe toward which all EBBPs/MSPs who have *never* been BBF members scurry frantically in the hope of "making the cut" for the next BBF family of potential cosmic, universal Hell-TV play-actors.

Chapter Eighteen: What Happens When We Die?
What Might Happen When We Die?

We might finally get told the truth about our *cosmic centrality* (CC) *after* we get rid of our problematic bodies and pass through a black-hole tunnel (an "Einstein-Rosen bridge" [ref. 12, p. 94]) with a bright light at the end of it (the same phenomenon as that observed and experienced by most out-of-body and near-death travelers).

At the end of the black hole, Einstein-Rosen tunnel, you might get greeted not only by all of your "deceased" loved ones but also by every other "deceased" (person/BBF member) who was ever born on Earth, including everyone who was famous *within* the world during their respective lifetime. You might also get greeted by all of the BBF members who have yet to be born and live on planet Earth-Hell.

So, you might well be greeted not only by your ancestors but also by your future, yet-unborn descendants as well as by Jesus of Nazareth and the other two parts of the Holy Trinity, Abraham Lincoln, Albert Einstein, etc. Once our bodies are dead, the infinite universe is no longer jealous of us (and we're no longer in the direct line of their fiery wrath) because we are no longer within the central focal point within the CRP, and therefore we no longer have the honor/distinction-burden/affliction of CC (cosmic centrality).

Once we've left our cumbersome, *intended-as-a-punishment* (by our cosmic enemies) bodies and traveled down, through and out of the black-hole tunnel(s), *we no longer have* any sharply defined, that is, *focal CC for any never-been-included-in-any-Big-Bang-Family-member-relegated-to-a-CRP List EBBP/MSPs* to be viciously jealous of. At that point, we're out of harm's way (the way we were when we resided on Earth-Hell).

When we die and are liberated from our burdensome and tiresome bodies (which *they* afflicted on us, "they" meaning our Heavenly enemies as adjectivally delineated in the previous paragraph), we finally get told that we were (infinitely) famous throughout the entirety of our (in most cases, boringly, non-*worldly*-famous) lives. And from this time after "death" onward, we will each be able to enjoy and appreciate our respective *permanent, eternal, they-can't-take-it-away-from-us, cosmic celebrity status* (CCS).

Chapter Nineteen: Why No Suffering in the Universe at Large

What about suffering (and pain) throughout the infinite, Heavenly universe?

Apart from life in the CRP, there is (and there can be) no suffering and no (unwanted) pain, because once you're rid of your nuisance of a body, if anyone tries to inflict any (unwanted) pain and/or suffering on you, all you need to do to avoid and escape from it is to willfully dissolve into infinitesimally small particle-ettes and thereby effectively cease to exist (temporarily) and thereby disappear from your enemies' midst/whereabouts and reappear/reconstitute yourself (your MSP) at some location too far away from them for them to know where you are or to be able to do anything harmful to you even if they could know where you were.

Chapter Twenty: Within the Next 5 to 100 Hundred Years

What might happen on Earth within the next 5 to 100 years and how we could expedite this progress.

Within the next 5 to 100 years or so, via brain stimulation, mind particle stimulation, reproducible out-of-body experiences, the Internet / World Wide Web, global positioning systems and other technologies, every MP that has ever been imprisoned within a human (or beloved, nonhuman, pet animal) body will be liberated therefrom and will have full and *equal* access to every other such MP's knowledge, financial wherewithal (actual money will become obsolete), *fame* and other "riches," including full past histories or biographies.

Hence, since all past, present and future earthlings will *share* all aspects of our wherewithal equally with all other earthlings, we will all be equally famous, rich, geniustic, etc., forever.

All of this *technologically facilitated* effective *mental telepathy* might prove reminiscent of the *non-technologically facilitated* mental telepathy of the reputed great healer (?) Edgar Cayce (1877–1945) who *may* have had access through subconscious but "direct communication with all other subconscious minds, and (was) capable of interpreting through his objective mind and imparting impressions received to other objective minds, gathering in this way all the knowledge possessed by *millions of other subconscious minds*" (ref. 13, p. 94).

The way in which we might be able to expedite these improvements in the "human" condition might be by deliberately anticipating them and cultivating the appropriate research and development.

Chapter Twenty-one: Everyone's Right to Euthanasia

Everyone's right to conscientiously-considered, painless, physician-assisted suicide.

Since each of us MPs who are imprisoned within a human body is doing a *favor* for the entire universe by (1) populating its current CRP and (2) entertaining it with our pain, suffering and boredom, we should have the globally acknowledged right and access whereby to painlessly and sufferinglessly *limit* how much of a favor, that is, for how long of a period of "life"-time we're willing to do the favor.

The specifics of how *conscientiously considered* painless, suffering-free suicide might be administered are beyond the scope of this essay, but will be addressed at a later time, in a follow-up essay. For now, suffice it to say that if any person consistently expresses a wish for such suicide for the sustained duration of three or more years' time, then that person should be granted access to the means whereby to fulfill that wish.

What's special or sufficient about a three-year waiting period? Well, Jesus of Nazareth, whether one believes Him to be God (as I do) or simply a Great Prophet, reportedly (ref. 1) spent only three years actively pursuing His teaching purpose here on Earth. If three years were time enough for Jesus to accomplish *His* grand purpose, then certainly that would be time enough for *anyone* else to accomplish *their* purpose here on Earth.

Furthermore, cases of extreme pain and/or suffering, such as might be entailed in effectively untreatable physical *or mental* illness, should be accommodated with death-wish fulfillment almost immediately or well before three years have elapsed. In these expedited cases of euthanasia, the prerequisite pain and/or suffering criteria should be measurable via *biometric* methods (ref. 2).

References

1. Crystal, D. *The Cambridge Encyclopedia*, Cambridge University Press, 1994: 590.
2. Blakeslee, S. Scanner pinpoints site of thought as brain sees or speaks. *The New York Times*, Tues., June 1, 1993: C-1 and C-3.

Chapter Twenty-two: A Preview of the Next 1 to 50 Years (with References)

A preview of remarkable advances that could and probably will occur between now (the present moment) and 50 years from now.

Part I
Improving Learning Abilities and Work Skills

Learning abilities and work skills could be significantly, if not dramatically, improved by stimulating the brain's pleasure centers (for example, the ventromedial frontal cortex and/or nucleus accumbens, probably on the left side of the brain) (refs. 1–18), possibly while at the same time stimulating superficial areas of the cerebral cortex (ref. 19) via a biofeedback loop (an external-to-the-brain circuit) (ref. 20) that would be designed in such a way as to deliver the pleasurable brain stimulation *whenever and only whenever* the student/worker were emitting brainwaves in the 12 to 42+ Hz (cycles per second) frequency range which indicates that the student/worker is engaging in high-level, such as learning- or work-related activities (refs. 21–24).

By doing this, the student/worker would get the rewarding, pleasurable brain stimulation (BS) *if and only if* she or he were studying and working diligently. Such an approach would be conducive to high-quality learning and work experiences.

Part II
Enabling Everyone on Earth to Know Everyone Else, Hence Be Globally Famous Celebrities

Reproducible out-of-body experiences (refs. 25–27) that would enable each person's MSP to leave their brain-body pair and travel out and among and interact with, and thereby come to know well, every other earthly person's MSP (hence, everyone would be famous) could be accomplished by stimulating certain areas of the brain together with or coupled with pleasurable BS that would be delivered *whenever and only whenever* the person's/s' brainwaves would indicate that he or she were having an out-of-body or "near-death" experience (ref. 20).

Since we already have thousands of people with brain-implanted electrodes (ref. 28), the ethical issues surrounding the implementation of BS have already been negotiated. Moreover, stimulating anywhere in the brain is now or is about to become something that can be done entirely surgically-noninvasively (without implanted electrodes) via *multiple*, external-to-the-brain-and-head, *electromagnetic* coil configurations (ref. 29) and/or externally focused ultrasound (refs. 30–32).

References

1. Spergel D. Is the universe finite or infinite? Discover presents (the) top 75 questions of science. Why is there something rather than nothing? *Discover*, Spring 2008: 6.
2. Powell C. S. Discover presents (the) top 75 questions of science. Why is there something rather than nothing? *Discover*, Spring 2008: 7.
3. Battersby S. Wanted: a theory of everything. *New Scientist*, April 30, 2005; 186 (2497): 30–34.
4. Chown M. The hypertime trap. *New Scientist*, October 13, 2007; 196 (2625): 36–39.
5. Kaku M. *Hyperspace: a Scientific Odyssey through Parallel Universes, Time Warps, and the Tenth Dimension*: Anchor Books, Doubleday 1995, copyright 1994 by Oxford University Press: 360 pp.; 109.
6. Mancini L. *How Everyone Could Be Rich, Famous, Etc.* Trafford Publishing, (March 8) 2006: 191.
7. Ref. 5, p. 87.
8. Ref. 2, pp. 3–9.
9. Kruesi L. Cosmology: 5 things you need to know. *Astronomy*, May 2007: 28–33.
10. Sears G. M., personal communication: 2001.
11. Crystal D. *The Cambridge Encyclopedia*: Cambridge University Press, 1994: 1003.
12. Ref. 5, pp. 224–226.
13. Sugrue T. *The Story of Edgar Cayce: There Is a River.* A.R.E. Press, Association for Research and Enlightenment Inc., revised edition, 1994: 384 pp.; 154.
14. Ref. 11, p. 290.
15. Blakeslee S. Scanner pinpoints site of thought as brain sees or speaks. *The New York Times*, Tues., June 1, 1993: C-1 and C-3.
16. Wilson J. Why We Laugh. *Popular Mechanics*, March 2003: 40–41.
17. Kinsley C. H., Kelly G. L. The maternal brain. *Scientific American*, Jan. 2006, 294 (1): 72–77.
18. Begley S. How the brain rewires itself. *Time, Mind & Body Special Issue*, Jan. 29, 2007: 169 (5): 72–79.
19. Osborne L. Savant for a day. *The New York Times Magazine*, June 22, 2003; (6): 38–41.
20. Butler S. R., Giaquinto S. Technical note: stimulation triggered automatically by electrophysiological events. Med & Biol Engng Comput, 1969; 7: 329–331.

21. Biello D. Searching for God in the brain. *Scientific American Mind*, October/November 2007; 18 (5): 38–43.
22. Voelker R. Ramping up rehabilitation research urged as a "public health imperative." JAMA, Nov. 16, 2005; 294 (19): 2413–2416.
23. Webb J. Bionic brains. *New Scientist*, May 28–June 3, 2005, 186 (2501): 32.
24. Fox D. Brainwave boogie-woogie. *New Scientist*, Dec. 24–Dec. 31, 2005; 188 (2531/2532): 50–51.
25. Kotler S. Extreme states. *Discover*, July 2005; 26 (7): 60–66.
26. Bosveld J. Soul search: can science ever decipher the secrets of the human soul? *Discover*, June 2007; special issue: 46–50.
27. Hoppe C. Controlling Epilepsy. *Scientific American Mind*, June/July 2006; 17 (3): 62–67.
28. Hall S. S. Brain pacemakers. *Technology Review: MIT's Magazine of Innovation*, Sept. 2001; 104 (7); 34–43.
29. George M. S. Stimulating the brain. *Scientific American*, special issue, Sept. 2003; 289 (3): 66–73.
30. Basu P. Sterilized by sound. *Discover*, May 2003: 13.
31. Davies J. Prophetic patent. *New Scientist*, April 23, 2005; 186 (2496): 31.
32. Hogan J., Fox B. Sony patent takes first step to real-life matrix. *New Scientist*, April 9, 2005; 186; 10.

How Everyone Will Be Rich, Famous, Painless, Deathless, Well Educated, Sexually Liberated, etc., Starting Sooner or Later

or:

How everyone could be rich, famous, pain and death free (immortal), well educated, sexually liberated, unselfish, healthy, free of anorexia, obesity, nausea, blindness, deafness, paralysis, insomnia, etc., via brain stimulation, neural prostheses, pacemakers, mind particle internet, circulation, cloning, genetic engineering, conscious computers, etc.

(completed on or before September 7, 2000)

Lewis S. Mancini

SUMMARY

Financial stress might be markedly diminished, while employment-related knowledge, productivity and adaptability might be augmented and diversified by a (preferably, surgically-) **noninvasive** brain stimulator, pacemaker or other kind of nervous-system-function modifier or neural prosthesis. This device or method would deliver highly enjoyable brain stimulation whenever and only whenever one were to engage in high-level (of complexity or of time-pressure-related intensity, economically appropriate and sellable) mental or physical activities, as indicated by physiological criteria. Sexual liberation might be achieved through brain-stimulation-mediated enhancement of (a) freely-chosen sexual impulses and suppression of (b) impulses one would prefer not to be compelled by. It may be possible to circulate everyone's **mind particle** (**MP**, the hypothesized biological and physiological basis of each person's individual mind or consciousness) through everybody's **brain-body pair** (i.e., body, including its associated brain), **BBP or, simply, BB,** so as to enable each consciousness to directly experience and, therefore, be empathetic toward every BB's needs and wants. No longer being confined within one particular BB, each individual mind (particle) would be unaffected by any given BB's death and might always be able to transfer, when necessary, to a viable BBP, thereby avoiding death. A less fascinating, less educationally valuable way to avoid dying would be to simply transfer oneself (i.e., one's mind particle) to a youthful clone of the original BBP one **was born into** each and any/every time the latest-issued clone might become too decrepitly old or otherwise impaired to continue living.

HOW EVERYONE WILL BE RICH, FAMOUS, PAINLESS, DEATHLESS, WELL EDUCATED, SEXUALLY LIBERATED, ETC., STARTING SOONER OR LATER

HOW EVERYONE WILL BE RICH AND WELL EDUCATED

A basic learning and work facilitative idea at a glance

The methodology about to be proposed might entail the use of (1) a stimulant modality such as ultrasound (US), electromagnetism (EM), beams of subatomic particles, ionized pharmacotherapeutic atoms, molecules and/or some other kind(s) of chemical/physical phenomena (perhaps, precisely) targeted, focused, conveyed and applied, possibly, entirely from outside of the fully intact head together with (2) a simple (possibly, biofeedback-based or driven) circuit, whereby a person would receive pleasurable stimulation (that would be pleasurable/rewarding by virtue of <u>which</u> particular neuroanatomic sites and stimulus parameter values might be used) of one or more reward-mediating pathway(s) or site(s) in the brain or elsewhere in the nervous system, if, and only if, whenever and only whenever, and for as long interval(s) of duration or period(s) of time and only for as long intervals of duration or period(s) of time as, that is, during or throughout any intervals of time during which the individual (stimulation recipient) were to engage in/were engaging in high-level (of complexity or intricacy) mental and/or physical processes and activity/ies, as evidenced by and driven by specific-to-process-or-activity physiological manifestations, that might serve as biofeedback signals (refs. 1–20). Then, consequently, that is, primarily, due to conditioning induced by the virtual or near-simultaneity of (a) these processes/activities with (b) rewarding experiences of pleasure/enjoyment, hence, by dint of instrumental chronological association, virtual synchronousness or pairing in time of stimulus (pleasurable stimulation) and response (high-level mental/physical process or activity/ies coupled with characteristic physiological emissions or manifestations), this/these activity/ies and process/es might be experienced as (being themselves, virtually intrinsically) intensely pleasurable and, therefore, interesting and likely to occur often and for long periods of time. The process or activity in question might be learning, reading, problem solving, memorizing, remembering, skilled manual or any other type(s) of work.

Electroencephalographic (EEG), electromyographic (EMG), magnetoencephalographic (MEG), specific-brain-function-indicative or functional magnetic resonance imaging (fMRI) and/or some other kind(s) of detectable, physiologically significant phenomena (that might serve as biofeedback), which are identifyingly characteristic of such activities/processes, would necessarily be the turn-on and stay-on signal(s) (in other words, the bio- or neuro-feedback signals) for the brain stimulator (BS) or (neural) pacemaker/pacesetter or prosthesis (BP or NP). A person could quickly become intensely interested in, highly motivated with respect to the associated details of any area or career-related direction and, thence, extensively knowledgeable in any area(s) of endeavor the person/stimulus-recipient were to choose. Being so productive, patient (defining "patient" as: "well enough equipped with pleasure in order to offset and undermine any frustration/impatience"), and versatile, a person would be unlikely to suffer from financial stress, poverty, hunger or other shortage or lack-of-money-related adversity.

It is conceivable that these effects could all be produced without involving any direct contact between the stimulator and the person being stimulated. In any case, not only might there be no surgery, pain or discomfort involved, but quite possibly, there would be no tactile sensations of any kind (1–2). In view of already-accomplished research and development (3–20), it can be hypothesized that a device that would utilize the principle suggested above could probably be developed in less than a year's time, at a cost of (conceivably, as little as) several million U.S. dollars, which would be a minuscule investment in comparison to the billions of dollars worth of increased human productivity and new, improved and diversified work skills it might facilitate.

A definition and fuller explanation

A useful definition of the word "**rich**," for this context, might be: (a) **financially able to fulfill all of one's biological needs** and (b) **able to lead one's life without experiencing financially-related anxiety or stress.**

It might be appropriate to use **biochemical** and/or **biophysical stimuli** (distinction explained below), focused and **applied** (preferably) **noninvasively** (at least, **surgically-noninvasively**; distinction explained below) by means of a relatively simple circuit, perhaps conceptually similar to one designed approximately 30 years ago (3). This device might usefully be designed in such ways that an individual would obtain A, that is, (preferably, **highly**) **pleasurable** or at least relaxing, soothing **stimulation** of one or more pathways, sites, areas or structures **in the brain or elsewhere in the nervous system** (for example, the vagus nerve) (4).

However, the stimulation delivery system would be designed in such ways that the individual would receive A, that is, **pleasurable (or, at least, comforting, relaxing and patience-conducive) brain or other neural stimulation if and only if, for as long a duration of time and only for as long a duration of time, whenever and only whenever** the individual (stimulation recipient) were to engage in B, that is, **high-level (of complexity or of time-pressure-related intensity) mental and/or physical activities and/or processes, as indicated by the evidence of C, that is, physiological or physiologically-related manifestations or signals (5–15) of high-level mental/ physical, activities/processes** that (at least consistently, intraindividually or consistently for each individual stimulation recipient) might be expected to be reliably, consistently and identifyingly manifested in the same detectable way(s) for each occurrence of the same activity or process. Any given high-level, mental/physical, activity/process might be expected to reliably, consistently and distinctively manifest itself in a detectably characteristic way, in terms of **functional magnetic resonance imaging (fMRI), computer-analyzed, spectral-analyzed electroencephalography (EEG)** or some other modality. Each occurrence of a different high-level, mental/physical, activity/ process might be expected to manifest itself reliably and consistently, yet distinctively differently with respect to any occurrence of a low-level (of complexity or of time-pressure-related intensity) mental/physical, activity/process, such as relaxing in front of a television.

The high-level, mental/physical activities/processes would be inherently (by dint of the functional nature of the brain and of the mind) coupled and concurrent with (sometimes, keenly observant, hence, highly perceptive and, at other times, deeply introspective, hence, highly intuitive) **sharply-focused attention** and virtually **uninterrupted, amply-sustained, thoroughly engrossed concentration**. Explained somewhat differently or, at least, more generally: each different kind of learning-related and/or work-related (generally, high-level) constructive and productive mental/physical, activity/process might reasonably be expected to have corresponding, relatively-uniquely-identifying (at least within any given stimulation recipient) activity-specific or process-specific, externally detectable and monitorable physiological or physiologically-related manifestations (5–15).

High-level mental/physical activities/processes (Bs) might include **learning** (especially, rapid-rate or high-speed learning), whether it be of abstract or of tangibly practical information and concepts, **reading, problem solving, memorizing, remembering** (that is, retrieving memories or already-learned and, to some degree at least, mastered items of knowledge and information that would already be fully processed, that is, encoded and stored in the brain and/or in the mind; please note distinction below), and **working** (especially, rapid-rate or high-speed working, in particular, effective and proficient application or implementation of high-level work skills and learned task-accomplishing behavior patterns). The physiological or physiologically-related manifestations, referred to above, that might be emitted (as stimulation-initiating and stimulation-sustaining signals) by the stimulation-receiving individual, might be expected to be instantaneously detectable, sustainable and recordable in real time.

For simplicity of discussion's sake, let us assume that it is physiologically unfeasible (if not impossible) to separate **B (high-level, mental/physical, activities/processes)** from **C (the measurable, quantifiable, physiological or physiologically-related manifestations** of high-level, mental/physical, activities/processes). Consequently, whenever a person might engage in B, the brain-stimulative/neural-prosthetic device would detect C-type manifestations and, therefore, begin to deliver A (**pleasurable stimulation**) to this particular person. So, the person would have a reliably strong motivation (i.e., pleasurable stimulation or pleasurable activation/excitation) toward indulgence in high-level, mental/physical, activities/processes.

Thus, **because the pleasure would be actuated**, driven and **maintained by**, and also nearly, if not entirely, **simultaneous with** the sustained-by-way-of-rewarding-brain-stimulation-initiating-and-sustaining **manifestations** (emitted by the stimulation-recipient) of the **high-level** activities/processes, **these activities/processes, themselves**, by virtue of chronological association (entailing near- or semi-simultaneity), **might be experienced as pleasurable** and, **therefore, interesting** and **likely to occur often and for long periods of time**. Thus, the **physiological/ physiologically-related manifestations** of these activities or processes would instrumentally serve as the **turn-on and stay-on signals** for the brain or neural stimulators. Accordingly, any person might be easily enabled to readily become intensely pleasured/ **pleasurized or gratified** by this person's mind/brain's attending to and processing even very minute details and very subtle conceptual nuances encompassed within any area of academic knowledge or skilled, occupational/ vocational, practical-expertise-requiring, employment-related implementation.

By dint of a mechanism such as the aforementioned one, any person could quickly become intensely pleasured with respect to, hence, strongly interested in and knowledgeable about virtually any kind of **high-paying** work skills (or hobbies of any kind) that the person might choose to become interested in and knowledgeable about. For example, suppose an individual is someone whose favorite pastimes are indoor housekeeping and outdoor gardening and there are no high-paying jobs available in either of these areas. Now suppose the person realizes there are many high-paying job openings in high-technology fields, such as biomedical engineering.

This individual might be able to start receiving A (pleasurable brain stimulation/neural pacemaking) whenever the person were to engage in attentive reading of biomedical-engineering-based/-related printed materials or otherwise perceptually accessible media, such as might be accessible, for example, via computer screen. Then, whenever the person were to engage in B (biomedical-engineering-relevant, high-level, mental/physical or mental/behavioral, activities/ processes), the brain stimulator would be sending out A (enjoyable brain stimulation) in response to its detection and monitoring of the person's emitted C (-type physiological/ physiologically-related manifestations) that might not even be directly known about (not the details, anyway), much less cared about by the person in question. From the would-be housekeeper/gardener's viewpoint, suddenly the study of high-technology subjects (such as biomedical engineering, etc.) might, almost mysteriously, take on the attraction of great fun. And better yet, the prospect of a high-paying job might almost as unexpectedly, suddenly become a genuine possibility that might expeditiously be fulfilled.

Hence, if a person were to have access to A (pleasure-inducing brain stimulation/neural-pacesetting) and circumstances (such as a shortage of money) leading to or amounting to a rationale for becoming knowledgeable and capable with respect to high-level, mental/behavioral, activities/processes, then the person's prospects of being or becoming able to do a good job while being employed in a **high-level-of-practical-expertise-entailing occupation** might be excellent. And regarding the preexisting hobbies and interests (in the hypothetical example being considered here) involving housekeeping and gardening, there would be no reason to anticipate a loss of enthusiasm in relation to these areas of endeavor, although they might get relegated to the roles of recreational diversions that would, nonetheless, continue to be enthusiastically partaken in (without brain stimulation or neural pacesetting being necessary or appropriate).

Almost any person being rewarded in this **high-level**-activity-process-**manifestation-dependent** way, with **interest-generating and patience-** (definable here as **relatively-low-intensity-pleasure-**) generating and accommodating, gratifying brain/neural stimulation, pacemaking, might readily and easily become deeply interested in and engrossedly patient with respect to both major concepts and numerous minute details of any (preferably, well-paying) field of endeavor. Consequently, the individual might readily and quickly become well educated about descriptive factual information, concepts and (as appropriate and financially necessary, high-speed implementation of specific) work skills, related to **any areas of endeavor** and **modes of practical application** that the individual might choose or that the ambient economy might demand or require, in accordance with whatever volumes of **supplies** of whatever **goods and services** might be readily **sellable and financially beneficial**.

The stimulation might consist in nerve cell/**neuronal activation** and/or **deactivation, excitation/arousal and/or inhibition/suppression**. It might be experienced by the **stimulation-recipient** as pleasureful by dint of a combination (a) which specific stimulative modalities media or means, (b) the magnitudes, values, sizes, or other quantifiable characteristics of stimulus parameters/variables and (c) which particular neuroanatomic target-sites would be used (4, 16–36).

Also, depending on which particular stimulation modalities, stimulus parameter values and target sites were to be used, possibly, no pain, tactile sensations, other perceptions (1, 2, 30), nor even appreciable effort (as explained below) would need to be involved or be unavoidable.

The effort, which generally (for most people, at least), seems to accompany or accompanies learning and (what we think of as) work(ing), may be thought of as goal-directed pain. The pleasurable facilitation of learning and work, as suggested herein, might be expected to minimize or even effectively eliminate the need for (this pain, which is the essence of) effort, perhaps by the mechanism of (a) **reciprocal inhibition** (37).

Reciprocal inhibition can be understood as follows: **excitatory stimulation of pleasure-subserving** or, in other words, pleasure-mediating anatomic sites might entail indirect inhibition of pain-subserving sites, by means of naturally-occurring inhibitory projections or pathways, extending from pleasure-subserving to pain-subserving sites. Analogously, pain-mediating sites might have naturally occurring (reciprocally) inhibitory pathways that extend to pleasure-subserving sites.

Another way or mechanism by which learning and working might be virtually **divested of** their painful/effortful aspects might be by (b) **simultaneous excitatory stimulation of pleasure pathways** or sites **and** (together or chronologically coupled with) **direct inhibitory stimulation of pain-subserving pathways or sites**.

Regardless of the mechanism(s) by which inhibition of learning-/work-associated (pain of) effort might be achieved, the phenomenon of such inhibition, entailing minimization or even virtual elimination of the need for expenditure of effort, might be referred to as the **de-effortization, deeffortization, deffortization, deffortation or defortation** of learning and work(ing).

Analogously, the phenomenon of learning and work being perfused with brain/neural-stimulation-mediated pleasure and being rendered reliably, highly enjoyable might be referred to as the **enjoyitization, enjoyitation or pleasurization** of learning and work.

By means of the pleasurization and relative defortation of learning and work, the human condition as we (most of us, anyway, seem to) know it, that is:

a)   we (each of us, individually) tend to experience much pleasure when we (ourselves) succeed at a learning- or work-related task and

b)   we tend to experience much pain or displeasure when we fail at a learning- or work-related task **might be changed** so that:

c)   we would still experience much pleasure (maybe more than we do already/currently) when we would succeed but, most significantly,

d)   we would, due to brain-stimulation-mediated inhibition of pain and suffering, experience virtually **no pain or displeasure** when we would fail.

Consequently, we (each of us, individually, and all of us, collectively) might tend to be elated by our successes without being dejected or demoralized by our failures. Since (c) and (together with) (d) correspond to a relatively optimistic outlook, it can be inferred that the human mental condition, as a whole, might be improved. Moreover, as can readily be surmised or inferred by comparing one's observations and impressions of the learning and work performance of optimists with that of pessimists, **the likelihood** (and frequency) **of success and quality of performance** might, comprehensibly, be expected to, respectively, **increase and improve**.

Depending on each (that is, any given) stimulation-recipient's own choice and exercise of free will, (pleasurization and defortation, hence) facilitation of mental/physical processes might be such as to only be of value to the stimulation-recipient as amusing recreational phenomena of the pastime or hobby type. However, more valuably, from a financial/economic standpoint, each recipient might better be persuaded that the facilitated mental/physical processes might more wisely be **highly relevant to the realm of gainful employment**.

Summarizing some points made above, an important point to keep in mind is that high-level mental/physical activities/processes might be determined to have (as their characteristically associated and each-individual-stimulation-recipient-identifying) physiological/physiologically related manifestations, some possibly more or less unique to each recipient, consistently recurring patterns of data-points (signals) of any of a number of different kinds of detectable, measurable and recordable modalities that might be used as stimulation-triggering and stimulation-maintaining/ sustaining signals (5-15).

A list of these modalities might include electroencephalography (EEG), quantitative, computerized or computer-frequency/spectral-analyzed EEG (CEEG), magnetoencephalography (MEG) (38), functional magnetic resonance scanning and imaging (fMRI), optical imaging (39), electromyography (EMG), thermography or thermographic radiant-heat-quantity-related infrared (IR) transducer-mediated imaging, (perhaps, three-dimensional) ultrasonography (US or 3DUS), Doppler ultrasonography, positron emission tomography (PET) scanning, single photon emission computed tomography (SPECT) scanning and others.

As learning-facilitative (LF), work-skills-acquisition-facilitative or work-skills-performance-facilitative (WF) brain/neural-stimulation/pacemaking/prosthetically-mediated turn-on and stay-on signals, these mental/physical, activity/process-identifyingly-associated/identifyingly-linked manifestations/characteristics/signals might be referred to as Learning/Work-Associated/-Linked Manifestations (LWAMs or LWLMs), Learning/Work-Associated/-Linked Characteristics (LWACs or LWLCs) or Learning/Working-Associated/-Linked Signals (LWASes or LWLSes) (5–15).

The following two quotations might seem to convincingly/plausibly validate the notions that such signals both exist and are medical-technologically accessible: on page 116 (of the reference source), (a) "A close correspondence exists between the appearance of a mental state or behavior and the activity of selected brain regions" and, on page 115, (b) "Neuroscience continues to associate specific brain structures with specific tasks" (40).

For this context, biochemical stimuli might be thought of as any stimuli that predominantly entail particles or particulate matter, such as one or more biological- or medical-effect-mediating, pharmacotherapeutic molecules, whole atoms, ions, etc. Whereas, **biophysical stimuli** might be thought of as any biological- or medical-effect-mediating stimuli that predominantly entail subatomic particles (e.g., electrons) or biological- or medical-effective-mediating wave(form)s of virtually any kind, such as sound, electromagnetic waves or quantum wavefunctions, such as can or might be retroreflected from a boundary/interface between a normal conductor and a superconductor (41).

However, in view of the quantum mechanical phenomenon of wave/particle duality of energy and matter, it is clear that biochemical and biophysical stimuli cannot validly be considered strictly, mutually exclusive or categorically distinct from each other.

The terms "**noninvasive**" and "**surgically-noninvasive**" **are not, in this context anyway, interchangeable**. For example, in proceeding toward the goal of explaining the appropriate distinction, let us consider transdermal permeation or percutaneous transmission of microminiature (42) robotic, electronic and/or pharmacologic/medicinal agents or "**nanobots**," (a term used by R. Kurzweil defined, in essence, paraphrastically, adaptedly herein) as explained in detail on pages 307 and 68, respectively, in the next-cited two references (39, 43) as "self-replicating" entities (robots built by means of nanotechnology) that are tiny enough for billions (or even trillions) of them to smoothly travel through a nanobot-recipient's circulatory system. Within such a recipient, the volumetrically minuscule anatomic environments affected might, as a consequence of being circulated through (by swarms of nanobots), undergo therapeutic (or otherwise adaptive-to-reference-frame-in-question) functional modification(s).

Thus, there might (at least, potentially) occur the phenomenon of **permeation** of any therapeutic or quality-of-life-enhancing entities, perhaps encompassing the full extent of traversing all the way from the periphery of an organism, through all intervening body-tissue types and from there proceeding deeply into all portions and ramifications of the circulatory system. Such an infiltrative phenomenon would surely qualify as an invasive process. However, because it would not rely on/or necessarily entail surgery of any type, it would not qualify as surgically invasive. On the contrary, it would validly be considered surgically noninvasive, despite being, in an overall, definitive sense, effectively invasive. More generally speaking, swallowing pills is an invasive treatment, but is clearly not surgically invasive.

In addition to permeation (44), a partial/noncomprehensive list of some other examples of **invasive, yet surgically noninvasive** means of potentially, possibly therapeutic/life-quality-improving **means of conveyance might include ingestion, inhalation and also injection via intermolecular, interatomic and/or even intra-atomic or subatomic interstices** so small that the injection recipient would not be aware of any skin being broken or any other disruption of the integrity or wholly intact condition of the tissue(s), anatomical structures or (components of) organs located between the therapeutic entry points on or in the recipient's body and the site(s) within the recipient that would appropriately be treated with the therapy.

The brain or neural stimuli and stimulation might consist in or be conveyed by means of (possibly, sharply focused) (a) electromagnetic modalities, such as transcranial magnetic (brain) stimulation (TMS) (4), (b) acoustic phenomena such as infrasound and ultrasound, (c) beam(s) or stream(s) of one or more individual (or kinds of) medicinal/pharmacologic molecules, atoms, ions or subatomic particles that might be forcefully powered, aimed at and precisely directed toward and into pleasure- (i.e., reward-) subserving sites in the brain or, conceivably, elsewhere in the nervous system or (d) other type(s) of (anatomic/neuroanatomic) target-location-specific therapeutic delivery system(s).

These systems might be readily capable of accessing precisely-focused-on, specifically-aimed-at target locations anywhere within the treatment-seeking subject or stimulation recipient. Such high degrees of localization or (high) resolution would possibly be, by current standards, highly nanotechnologized (43, 45) or microminiaturized. They might be circumscribed within very small areas, contained in minute volumes, measuring perhaps a tenth of a cubic millimeter or even some (tiny) fraction thereof (46).

The stimulus-focusing and stimulus-conveying/delivering apparatus might utilize waveform superposition principles (constructive and/or destructive interference) and/or mechanisms adapted and derivative from, or even actually entailing lasers, masers, phased arrays, tomography, holography, nuclear magnetic resonance, NMR (or, in terms of its medical applications, magnetic resonance imaging, MRI), superconductors, fiber optics, analog or digital circuits and waveform/signal-processing or other (perhaps, yet-to-be-conceived-and-devised) technologies.

The stimulation-mediating device would actually be a brain-, neural- or neurophysiological stimulator, pacemaker, pacesetter or other kind of nervous-system-activity-or-function-modulator, -modifier or -prosthesis, as explained presently. **Neural pacemakers** may be construed, mechanismically, as conveyors and deliverers of brain or neural stimulation, which is generated and applied in such ways as to engender some sort of pace, rhythm or pattern of response or responsiveness within or among cells (or their components) that can be maintained at some level(s) of balance, equilibrium or regulated interplay between activation and deactivation.

In view of many conceivable applications fitting the criteria of both brain/neural stimulation and brain/neural pacemaking/pacesetting, **all four abbreviated terms: (1) BS, (2) NS, (3) BP and (4) NP, may be considered practically equivalent and used interchangeably.** And these four categories might, collectively, **be designated as neural prostheses or neuroprostheses (NPs). The acronym BSNP can be used to signify any one, more than one, or even all four categories.** The NP part of BSNP may be interpreted as signifying (a) neural prostheses in general or (b) neural pacemaker-type, neural prostheses in particular, depending on which interpretation would seem more appropriate for any context being considered.

Consistent with an emphasis on how to improve and diversify learning abilities and work skills, neuroprostheses might (most readily and appropriately for this particular context) be construed as being **primarily intended to alleviate or minimize various learning disabilities**, attention deficits or deficit disorders, depressive, anxiety and other psychiatric disorders or illnesses, **as well as neurological or neuropsychiatric limitations** and deficiencies or **mere circumstantial, educational shortfalls** in relation to actual or potential high levels of technological prowess, as might be prerequisitely associated with high-level/high-paying employment opportunities.

A reason why a combination of (a) acoustic/sonic and (b) electromagnetic stimulation might be found to be adequate for many herein-suggested applications might be a mutually complementary (additive, potentiating or synergistic) interaction between the relatively high-level **focusing ability of ultrasound** (47–49) and the relatively high-level **penetrating power of magnetic stimulation** (4, 48, 49). The high level of magnetic penetration is consistent with the finding that magnetic fields are not impeded in their passage through (the skull) bone or through soft tissues.

Ultrasound (when appropriate and adequate magnitudes or values of stimulus parameters are used) might exert its stimulative effect(s) primarily by mechanically stretching, deforming and pulling open channels, pores or passageways in the neuronal/nerve-cell membranes and thereby increasing neuronal permeability and facilitating neural-activity-mediating ionic, molecular, or, in any case, particle-transporting and neurophysiological-activity-modulating currents, flowing inward toward, actually into and outward from/out of the neuron(s) (49).

And magnetic stimulation (when appropriate and adequate values of stimulus parameters are used) might exert its stimulative effect(s) primarily by creating or engendering (generally, ionic) current(s), flowing into and out from the neuron(s) (49). Thus, the ultrasonically-mediated stretching-widely-open of channels in the neuronal membrane(s) might facilitate and catalytically augment the magnetically-induced ionic currents that would increase neuronal activity, excitability and responsiveness in such ways as to facilitate brain, neural and neuronal stimulation and, thereby, readily implement the proposed mechanisms of learning abilities and work skills enhancement.

Consistent with the idea of using dual-modality, electromagnetic/ultrasonic stimulation would be the use of simultaneous, hence, **superimposed electromagnetic** and **ultrasonic fields,** perhaps with both being of the same frequency (46). Then, "the ultrasound field could be used to define the region to be stimulated, with the electromagnetic field providing a mechanism for stimulation, at reasonable field strength of each" as expressed by Robert A. Spangler (49).

One way (undoubtedly not the only way) of bringing about precisely-focused, high-resolution, reversible, noninjurious, functionally-modifying BSNP, might be achieved by combining (a) the basic approach delineated by the late William J. Fry in his 1968 paper (46) with (b) the basic approach entailed in "time-reversed acoustics" (41), including, perhaps, the use of **"time-reversal"** (i.e., **sequence-reversal,** stimuli-absorbing, stimuli-reflecting) **mirrors** (TRMs), as noted by Mathias Fink (41, 50–53).

Acoustic time-reversal mirrors are explained on page 92 (41), essentially as follows: "... a source emits sound waves. ... Each transducer in a mirror array detects the sound arriving at its location and feeds the signal to a computer ... each transducer plays back its sound signal in reverse in synchrony with the other transducers. The original wave is recreated, but traveling backward, retracing its passage back through the medium ... refocusing on the original source point."

An observation that might be of particular relevance, as expressed by Fink on page 97 (41) is the following: "Porous bone in the skull presents an energy-sapping challenge to focusing ultrasound waves on a brain tumor to heat and destroy it. A time-reversal mirror with a modified playback algorithm can nonetheless focus ultrasound through skull bone onto a small target."

It might be valid to infer from this observation that if one were to use somewhat different values of ultrasonic stimulation parameters/variables than those that might be useful in destroying a tumor, one might be able to (nondestructively, noninjuriously, and reversibly) functionally stimulate a small target area or volume in the living brain, inside of the fully intact head (meninges, skull, scalp, etc.).

Moreover, since time-reversal techniques may also be useful with **electromagnetic waves** (and quantum electron wavefunctions), they (T-R techniques) might offer a feasible way to implement W. J. Fry's idea of **combining ultrasonic and electromagnetic phenomena** to bring about "electrical stimulation of brain localized without probes." The crux of the combinative idea is expressed in the 1968 paper on page 919 (46) in this way: "The basic principle is partial rectification (in the focal region of an ultrasonic field) of the alternating current that flows in response to an externally applied electric field (of the same frequency). Since the magnitude of the electrical conductivity of the tissue varies with temperature, adiabatic temperature changes (produced by the acoustic disturbance) cause a periodic variation in conductivity that results in a net unidirectional transfer of charge."

It is conceivable that some adaptation of a combination of the insights of Fink with those of Fry might possibly lead to an acoustic/electromagnetic method capable of (a) detecting, monitoring and effectively utilizing learning-work(ing)-associated neurophysiological manifestations/neurophysiologically-related signals (LWAMs/LWASes) and (b) noninvasively stimulating reward- (i.e., pleasure-) mediating structures/pathways (16–36, 53–54) in the brain in notably effective, learning- and work-facilitative (LWF) ways.

A "**saser**" (**acoustically analogous to a laser**) might be characterized as "bright sound" or "a laser that's made from sound," expressed here in an acronym form that abbreviatedly represents the analogous-to-laser concept of "sound amplification by stimulated emission of (sonic) radiation" (55). Sasers are currently in early experimental and developmental stages. A diverse assortment of designs are in the process of being conceived and implemented in preliminary ways.

Component parts, media, and operative phenomena include sound-transmissible crystals, tiny blocks of glass or ruby (aluminum oxide containing sparsely distributed chromium ions), laser-beam-induced initiation of sound amplification, piezoelectric transducers that convert fluctuating voltages into high-frequency vibrations, semiconductors such as gallium arsenide consisting in laminated structural designs, high-energy-particle-bombarded pieces of silicon, water-filled vessels containing billions of tiny electrolysis-engendered gas bubbles that are squeezed by being subjected to electric fields or having their containers' sides squashed, **phonons** (high-frequency sound waves) or resonating lower acoustic frequencies. The goal of **amplified** (already accomplished to the extent of a factor of at least 30 "or so"), highly directional beams of sound waves is construed in terms of potential practical applications, as exemplified forthwith.

A list of intended and anticipated implementations of sasers might validly include powerful acoustic microscopes, defect-detecting, thickness-measuring and quality-assessing probes and sensors designed for use with respect to various composite materials, microprocessors/computer processing units (CPUs), devices that might increase signal-to-noise ratios (SNRs) in diversely constituted circuits by means of quietening the noise inherent in virtually all electrical circuits and, possibly, relevantly to the prospects of **detecting** and elucidating characteristics of **mind particles** (explained below), sasers that might be used in the capacity of ultra-sensitive particle detectors that would be analogous to photomultipliers.

Conceivable adaptations of sasers might also be useful when applied together with lasers as the respective acoustical/sonic and electromagnetic components of the aforementioned combinative paradigm entailing time-reversal (most relevantly in this context, stimulus/response pattern or sequence-reversal) (41) and probe-free, electrical brain stimulation/neural-function-modulating/pacemaking (46) technologies.

However, whether or not this or any other particular kind of approach should prove workable, it stands to reason that there probably, if not **undoubtedly**, could be **many more than just one way** of effectively accomplishing this potentially financially/economically valuable goal.

For example, a purely electromagnetic methodology or a purely acoustic/sonic approach might be adequately effective to accomplish significant or substantial learning/educational and work skills improvement and facilitation if suitable and adequate values of stimulus parameters are focused on / aimed at / targeted to pleasure (or "reward") centers possibly located in one or more of the following neuroanatomic areas, regions, pathways, sites, or structures (16): the medial forebrain bundle, some lateral hypothalamic and some limbic sites/structures, the cingulate cortex, anterior thalamus, hippocampus, amygdala, caudate, septal area, nucleus accumbens septi, forebrain/prosencephalon, midbrain/mesencephalon, medulla, entorhinal, retrosplenial and/or cingulated jutallocortex, or cerebellum (17, 18).

In particular, vagal nerve stimulation (19), septal stimulation (20), (whereby "patients brightened, looked more alert, seemed more attentive to their environment … could calculate more rapidly and, generally, more accurately (with) memory and recall enhanced or unchanged"), brain sites involved in the perception of speech (whereby auditory/vocal hallucinations, in a majority of schizophrenics tested, were diminished or alleviated (21, 22) and functional-MRI-indicated, simultaneous activation of sites in the prefrontal lobes together with sites in the parahippocampal cortex (11) might seem promising as potential memory, learning and work-skills improvement/facilitation, neuroanatomic-subserving foci (i.e., focuses)/loci.

Some other approaches/methodologies that might also prove useful/helpful in terms of brain stimulation/neural modulation/pacemaking might include some adaptations and applications of one or more of the following potential or already existing technologies: (a) contact-lens-size (56) or, conceivably, even (much) smaller (57–59) **electrodes**, merely placed in appropriate-for-specific-application position(s) on (the outsides of) patients', students' or workers' heads, in order to serve as brain-signal (56) and/or brain-stimulation mediators, (b) neural implants (39, 43, 60), such as nanobot-based implants that might circulate throughout the recipient's body (including the body's associated and component brain), **brain-computer-interfacing (BCI) with functional electrical**/electromagnetically-induced, physiological **stimulation (FES)** (61) being engendered thereby, (c) "neuron transistors" (such as adaptations of field effect transistors, FETs) (39, 43, 62–67) that "noninvasively allow communication between electronics and biological neurons" (43), (d) non-contact-requiring ultrasound (or other acoustic phenomena) possibly entailing "layers of material," added on to a sound emitter, "thereby matching (the emitter's) impedance to that of air" (68), (e) ultraviolet, visible or infrared laser-based or otherwise electromagnetically mediated electrical current or electron beams, possibly entailing one-molecule transistors (69), small-number-of-atoms- (e.g., 58 atoms) nanotechnologically comprised, ratchet-like motors (68) conceivably coupled with quantum tunneling, defined on page 313 (39) and electron-based quantum ratchets (70) that might constitute a basis whereby to eliminate the need for "a nightmare of connecting wires" and thereby implement a far-reaching realm of "wireless electronics."

What about situations in which rewarding brain/neural stimulation might actually be distracting, hence, obstructive to learning and working? Depending on (a) values of stimulus parameters, (b) the stimulation modality/ies and (c) which particular neuroanatomic "reward" or pleasure-subserving pathways, "centers" or sites might be used, pleasurable brain/neural stimulation might actually distract or pull the attention of the patient, recipient or aspiring student/worker/ worker-in-training away from the subject matter to be learned and mastered, thereby impairing rather than facilitating learning, working and other kinds of constructive performance and processes, such as hobby-related endeavors.

Two (but not necessarily the only) possibly effective solutions to this distraction/attention-undermining problem might be 1) to **use** some **other, different** stimulus characteristics or parameter values, some different stimulus modality/ies or means, and/or different neuroanatomic target sites and (2) to stagger, straddle or **alternate** (relatively brief) pleasure-stimulation periods with (relatively brief) learning/work-skills acquisition or application/performance periods.

By chronologically alternating (a) relatively brief (for example, five-minute-long) learning/ work/performance periods, while their constructive, high-quality occurrence would be virtually simultaneously substantiated and signaled by mental/physical-process-indicative manifestations with (b) relatively brief (for example, one-minute-long) pleasureful brain stimulation-delivery-reception periods, then, even if the stimulation were somehow or in some way(s) intrinsically distracting to learning, working or other modes of constructive activities (e.g., sports, hobbies), nonetheless, due to (the learning/working-performance-facilitation-stimulation system's entailing the conflict avoidance feature of the separation in time of the (a) learning/work/skills-performance period(s) from (b) the gratifying stimulation periods, the pleasure or "reward" periods might nevertheless be expected to motivatingly facilitate and improve learning, working and skills performance.

Conceivably, there might be a potentially feasible prospect of instantaneously fast, **effort-free, pleasure-free and even conscious-awareness-free** learning, working and other kinds of performance facilitation. Moreover, sooner or later, any feasible instantaneous/ultra-fast, altogether pleasure-free, pain-free, effort-free and subconscious or unconscious methods of **transferring, encoding and storing** enormous quantities of knowledge and information in virtually any brain or mind (possible nature of distinction explained below) might be discovered and implemented. But in the meantime, relatively defortized, pleasurized, rapid (but not necessarily instantaneous) and conscious-awareness-entailing or -necessitating (as opposed to consciousness-bypassing) learning/working/skills-performance-enhancing and facilitative methods might be significantly valuable.

Even without the discoveries or inventions of (a) **instantaneously-effectible**, conceivably (neurophysiologically) encodable/decodable **learning, work**-skills acquisition and skills **performance**-enhancement **facilitation (IELWPF) and** possibly **even without** (b) conscious-awareness-/conscious-mind-/**consciousness-bypassing learning, work and performance facilitation (CBLWPF), but merely with (c) enjoyitizing/pleasurizing, learning/work/ performance-enhancing-facilitation (EPLWPF) and (d) de-effortizing learning-working-performance-enhancing facilitation (DELWPF)**, the following scenario might be possible.

By means of EPLWPF and DELWPF (together specifiable as EPDE-LWPF), we might be enabled to be/become virtually instantly strongly interested **in and** quickly, **broadly,** maybe even **comprehensively knowledgeable about and proficient at the application of skills associated with or with respect to whichever areas of endeavor we might choose (with each of us making our own choices** and **decisions)** to be gratifyingly, prosperously and capably employed in terms of.

If and when we might ever have the benefits of all four kinds of LWPF (a–d, above), then, as **one** aggregate, fourfold kind of LWPF, it could be designated as **instantaneous, consciousness-bypassing, enjoyitizing/pleasurizing, de-effortizing learning, working** and other kinds of **performance enhancement** and **facilitation** (ICBEPDE-LWPF).

However, if the relatively readily implementable basic methodology of (enjoyitizing, pleasurizing, and de-effortizing) EPDE-LWPF, essentially as explained herein (above), were to be taken to fruition (under optimal research and development-associated circumstances, this goal might be accomplishable in as little as a year or two) (1–148), then any EPDE-LWPF recipient might readily be enabled to expeditiously become not only (conceivably) very well educated (maybe, in some cases, at least, encyclopedically so), but also (in connection with readily becoming sustainedly and reliably, employment-relevantly productive and versatile) unlikely to suffer from deprivation of material necessities or from financial stress, poverty, chronic hunger or any other shortage or lack-of-money-related adversity. Hence, any such EPDE-LWPF recipient/ beneficiary might be expected to be rich, according to the definition given above. A noninvasive learning/working-facilitative, brain-stimulation-neuroprosthetic-pacemaker/pacesetter system could conceivably come into practically useful being in as little as 1 to 5 years, in connection with and as a consequence of well-conceived and well-carried-out research and development, at a relatively inexpensive cost of several million dollars (1–148).

HOW EVERYONE WILL BE FAMOUS

Rene Descartes's renowned expression, "I think, therefore, I am," considered together with the realization that some **diseases** such as Alzheimer's, etc., **might deprive us of our ability to think**, might engender a modified version of his observation that might be expressible as: I experience, therefore, I am.

There is something about (or some aspect of) you, me and every other sentient, conscious individual which, on a lifelong basis, stays the same and maintains each of us as the selfsame individual throughout our entire respective lifetime. Is it (a) our genes? (b) our (cumulative) memories? (c) our personality/ies? or (d) some part of our physical composition or makeup, such as some irreplaceable cell(s), molecule(s), atom(s), subatomic particles, etc.? Is it one or more of (a) through (d) that maintain(s) each of us as our respective identities?

Although identical twins (and clones of any individual organism) are understood and considered to have the same genes, it is clear they are not one and the same individual or consciousness. This realization would seem to rule out genes as the lifelong identity-defining and identity-maintaining essence of each conscious individual.

It can be contended that "identical twins or biological clones are not 'the same person or people,' because they have different memories" (71). A thought experiment might elucidate the role (if there is one) of memories with respect to the identity of any given individual.

Suppose we have, as volunteer experimental subjects, two identical twin siblings, who can be designated as person A and person B or, simply, A and B, respectively. Now, suppose we, as a species, at some point in the future, have access to memory-erasing and memory-decoding, -encoding and memory-implanting equipment or devices. Suppose we were to erase each and every one of twin A's memories and implant in A's brain each and every one of twin B's memories (after having deciphered or decoded them carefully, noninjuriously and non-disruptively to B and then re-encoded and implanted all of them into A's brain).

Consequently, twins A and B would have all of the same memories (and no different ones). At this point, would A and B become one and the same person, mind or consciousness? No, of course not. A and B, at this point, would effectively have had all of the same experiences (that is, they would have all of the same records of experiences or memory traces, **"engrams"** [72] **or encoded histories of experiences**) in their respective brains/minds. But would having (records of) all the same experiences render these two **experiencers** one and the **same experiencer** (i.e., experienc**ER**)?

A distinction might appropriately be made here between (a) what gets experienced (i.e., experience**D**) and (b) who does the experiencing (i.e., experienc**ING**). So, despite having all the same experiences (records of experience), that is, all of the same memories (memory traces), A and B remain two separate and distinct experiencers. Hence, a difference of experiences (or memory contents) would not seem capable of explaining a difference of experiencers, that is, a difference of their identities. In other words, what gets experienced does not, in any way, specify who experienced it.

Twins A and B might be expected to have similar, but not identical, personality traits or features. Next, suppose we, as a species, at some point in the future, have access to personality-trait-modifying or personality-feature-erasing, -decoding, -encoding and -activating equipment.

Suppose we were to modify and/or erase each of A's personality features and put in their places or substitute each of B's corresponding features.

Consequently, twins A and B might have so perfectly identical personality profiles that the behavior of the two would be virtually indistinguishable from the standpoint(s) of any observer(s). Would two individuals' being characterizable as having identical or indistinguishably-similar personality profiles render them one and the same consciousness, mind or identity of experiencer? Or would they simply remain as two separate consciousnesses that would merely behave virtually identically in all observable situational contexts? The latter possibility would, intuitively, seem far more probable and credible than the former.

Thus, two individuals, even if (a) they are genetically the same (as in cases of identical twins), (b) even if they were to have identical records of experience, that is, identical cumulative memory contents and (c) identical personality traits and profiles, would still exist as two separate and distinct minds, conscious entities and experiencers. Hence, neither genetics nor memories nor personality features/profiles would seem capable of uniquely describing, specifying and constituting any given mind or consciousness.

The proposed answer to the question of what it is that constitutes the unchanging essence of each individual's identity might be that what maintains you, me and every other individual as the same individual throughout our entire, respective lifetime(s) is that the very same experiencing entity, that is, the same biochemically-/biophysically-based, discrete consciousness or, equivalently but expressed slightly differently, the potentially tangible, detectable, isolatable and implantable same experiencer (a thing or an object) will/would/does experience your, my or any other individual's respective life at age 110 (assuming, for discussion's sake, such prodigious longevity) as did experience it at age three (as well as at any other age, within or outside of the three to 110-year-age range).

You, I and any/every other individual might encounter some difficulty or doubts in responding affirmatively to the question, "Is the way I experience the world around me the way it actually, really is?" One might readily respond, essentially as follows: "Regardless of the accuracy/validity or inaccuracy/invalidity of how I perceive and interpret the 'world'/environment around me, which could, conceivably, be merely an exercise in virtual reality (73–76), one that entirely conceals from my perception an underlying real reality, in which I am merely a lifelong experimental subject, nevertheless, there are two things that I do feel relatively, confidently affirmative about and feel I do know, virtually for sure, are: (1) I am experiencing something; this life is something and (2) I am experiencing this something as a single, essentially separate, even solitary, individual, unitary entity, that is, a consciousness, who is aware of everything I experience. Or viewed and expressed somewhat differently, I am experiencing this, my life, as an individuated, undividable experiencer or indivisible consciousness."

But if you, I and anyone/everyone else is an indivisible experiencer, biomedically (i.e., biochemically/biophysically) bonded to or otherwise confined within a brain and its associated or corresponding body (abbreviatable as a **BB or BBP**, intended to signify a **brain-body pair**), what then is the biological or physiological basis of this undividable experiencer which is (the unitary) you?

Using the most powerful microscopes and medical imaging equipment and methods currently available to the human race, even if coupled with the most painstaking analyses and the most perceptive observations, it might (or might not) be readily possible to detect and discern a singular, discrete, unitary biological, physiological entity which is your, my or anyone else's indivisible experiencer or undividable consciousness.

One possible reason why we might not be able to visualize or isolate any single, unitary, undividable experiencer-consciousness (which is you, me or any other specific individual) might be that the biological, physiological entity in question might simply be too small to be visualized and isolated at this point in time, even when the best currently existing equipment is used.

Let us suppose that, at some point in the future, perhaps by dint of more powerful microscopes and imaging modalities than any currently available, it will become possible to actually see and then isolate each individual's indivisible experiencer-consciousness or, same-meaningly, mind, which, conceivably, may travel to any point(s) in the individual's body, but may spend the majority of its time, in accordance with what we seem to feel, in the brain. Such a visible, isolatable chemical/physical entity might be designated as a mind-particle-experiencer-consciousness or, alternatively, a mind-particle-consciousness-experiencer, abbreviatable as an MP, MPEC or MPEX.

In essence, **you are your MPEC**. The person you see in the mirror (any time you look) is merely the brain-body pair (BBP or, simply, BB) within which your MPEC, hence, your mind, is bonded, bound and confined by dint of implantation therein during gestation. A brain-body pair might, alternatively and equally validly, be designated as a body-brain pair (a body, including its component brain). The choice of the former designation (brain-body pair) rather than the latter is purely arbitrary.

Daniel C. Dennett, in his book *Consciousness Explained* (77), on pages 101 and 102, comes very close to conceptualizing this notion of an actual, discrete particle constituting the biological basis, the necessary and sufficient essence of each one of our respective, unique consciousnesses, each of which is a person's respective mind or consciousness, in Dennett's statement, "For most practical purposes, we can consider the point of view of a particular conscious subject to be just that: a point moving through space ..." The concept he suggests most clearly coincides with the herein-proposed hypothesis of a discrete mind-particle-experiencer-consciousness or, alternatively, a mind-particle-consciousness-experiencer (MP, MPEC, MPEX, or **MP-CONEX**) if the MP is construed to be a point-particle or point-sized, point-shaped (perhaps spherical) particle.

Also consistent with the hypothesis of a single, indivisible (mind) particle being the biochemical/biophysical basis of each person's individual, unique mind (or consciousness) is this quoted observation:

"... there is room for only one thing in the spotlight of attention at any one moment," which appears on page 32 in "States of mind ..." (14).

Such a mind or consciousness-particle might routinely, pervasively and information-gatheringly/-sharingly/-disseminatingly and -integratingly interact (by instantaneously making and then breaking huge numbers of chemical/physical bonds), directly or indirectly, with any and all chemicals constituting or otherwise contained within or passing through the human body, including, probably predominantly, chemicals within the body's component brain.

Subsequently, on page 430 in Dennett's book (77), reference is made to a material substance consisting in "some ... special group of atoms in your brain" as the hypothetical essence of any given person's consciousness, mind or identity. A reasoned response to this hypothesis might be that no one's individual/individuated consciousness, identity or mind could possibly consist in a **group** (?) of **any** thing or of any substance or kind. Since each person's identity, consciousness or mind seems to be just one entity, the biomedical, physical/chemical basis of individual identity (each individual's) can correspondingly, accordingly and rationally be expected to be just one physical thing/chemical thing and cannot plausibly consist in any kind of group (of anything). This one physical thing (or one chemical thing) cannot be, for example, a single atom, because even one atom is made up of a number, greater than one, of discrete, separate (subatomic) component parts or particles.

Regardless of whether or not the adult brain engages in the process of more or less continuously making new cells to replace older, dying ones (78–81), it stands to reason that if a group (of atoms or a group of any other kind/s of entities) is precluded from being the biophysical/biochemical basis of each or of any individual person's indivisible, unitary consciousness, then it also stands to reason that any group of brain cells (whether periodically-replaced or permanent on a lifelong basis) would also be precluded as the chemical/physical basis, that is, the necessary and sufficient, defining essence of each or any individual's singular, undividable consciousness, by dint of the same fundamental reason.

In particular, it does not seem logical or plausible that a group of any thing(s) (for example, a group of subatomic particles such as collectively constitute a single atom, a group of atoms, a group of molecules, of brain cells or a group of any other discrete entities) could possibly be the definitive physical-chemical essence/basis of any unitary, indivisible, singular phenomenon, such as each person's separate, unique, undividable and unitary mind (mind-particle-experiencer, MPEX) or consciousness (mind-particle-experiencer-consciousness, MPEC). This would seem to be true notwithstanding some very interesting and possibly correct (but outside the scope of this particular context) split-brain-studies-derived suggestions of there being at least two (separable) minds per BBP (82–84).

However, any given person's mind, or, definitively, any particular person's mental and physical/chemical identity could plausibly consist in any one specific, single, indivisible, subatomic particle contained in any one specific atom. But if this were true, then there might plausibly be trillions of trillions of minds (i.e., subatomic particles = consciousnesses) contained within each individual brain or brain-body (pair).

Considering the possibility that this idea (of trillions of individual, undividable subatomic particulate conscious minds inhabiting each living BBP) might be correct, let us conjecture that each one of the **subatomic-particulate-consciousnesses (SPCs)** might have some amount of free will and some amount of implementable power with respect to what the host brain-body pair as a whole says and does.

However, without having any particular reason to think otherwise, let us assume that each one of the trillions of SPCs has the same amount of free will and enforceable power over the BB (as a whole) as every other SPC does. And even though each SPC might (erroneously) believe itself to be the sole occupant/inhabitant of the BB in question, the fact would remain that it (any given SPC) would have no more (or less) power over the BB's speech and behavior than any other SPC residing therein.

Let us assume, for discussion's sake, that the amount of free will and enforceable power of each individual SPC with respect to the multi-trillion-mind-population-widely-shared BBP would be analogous and effectively proportional to the power of the one vote that each citizen in a democracy has the recognized right and ability to implement or cast in an election.

Under such analogous circumstances, each SPC would have no more ability to control or predict what the host BB would do or say (in relation to itself, in relation to other animated, apparently living BBs, or in relation to apparently inanimate, non-living objects) than an averagely empowered, individual citizen living in a democratic nation would have any ability to effectively dictate, control or predict events or news of an intra-national or international nature. Hence, each SPC would be continuously, somewhat surprised by the things its native BB would say and do, similarly as each of us forward-looking citizens is, perhaps regularly, somewhat surprised by what we read in each consecutive day's newspapers.

I will readily admit that I am occasionally surprised by the (sometimes embarrassing) things I say and do. But if I were merely one of trillions of conscious, one-vote-apiece-wielding occupants of what I consider to be "my" brain-body, with each of us (trillions of residents) contained herein (in this "my" BBP), mistakenly, feeling we are the sole inhabitant, then, I would expect to be regularly surprised by what my (host) BB says and does, continually rather than only occasionally.

By reason of the observation that I am only occasionally (rather than regularly) astonished by my words and/or actions, there would seem to be a strong suggestion that I (my mind or MP) truly am (or is alone), as the sole conscious determinant and, consistently (or, at least, not inconsistently), the sole conscious occupant of this brain-body pair wherein I reside. So, it may seem logical and (by dint of entailing a simplifying assumption or premise) sensible to return to the notion that each BBP is occupied by only one mind/consciousness, despite the (possibly correct) split-brain-studies-based suggestions of at least two minds/conscious entities per BB. So, let us proceed with this one MPEX per BBP assumption.

Explaining the situation somewhat differently, despite the often convincing, or at least strongly persuasive split-brain-study-derived notions of two conscious entities residing in each brain, if only for simplicity of discussion's sake, let us, in this context at least, embrace a one conscious entity (MP) per BBP assumption.

As delineated above, it would seem somewhat implausible to suppose that the chemical/physical basis of anyone's mind is merely one particular specimen of the tiniest undividably elemental category of subatomic particles of one particular atom contained within the BBP in which any particular mind happens to dwell or be bonded within.

It would seem more plausible to postulate that the biological-physiological, chemical-physical (hypothetically, perhaps, too-small-to-be-seen-or-imaged-with-existing-equipment) basis of the conscious mind is or might be an extremely tiny particle of a possibly as-yet-undetected and undiscovered kind of matter, that is, a conscious, experiencing kind of matter, in particular, a mind particle (experiencer-consciousness or consciousness-experiencer, MP-CONEX), as opposed to and different from possibly, relatively much more plentiful, presumably (but not unequivocally) non-conscious, non-aware, non-experiencing particulate matter that is composed of one or more particles, such as electrons, protons, atoms, ions, molecules, etc. Such seemingly non-experiencing, non-aware, non-conscious matter seems to be the mainstay of contemplation, theorization, experimentation and practical application as undertaken and implemented by an apparently vast majority of all kinds of scientists. It would seem readily conceivable that conscious matter, such as indivisible mind particles, might have properties that are significantly different from particles of presumably (outwardly, at least) non-conscious, non-self-aware, non-environmentally-aware matter, such as (presumably, that is, inferentially based on the preceding thought experiment entailing the somewhat implausible possibility of trillions of conscious, experiencing, self-aware, subatomic particles within each living brain/BBP) neutrons, protons, etc.

WOULD A CLONE OF YOU BE ANOTHER YOU?

If the undividable experiencer-consciousness or MPEC bonded within or inhabiting any particular clone of your BB were not your MPEC or MP-CONEX, then the clone would not be you. Instead, it would merely be a younger, virtual twin of you. If, on the other hand, we were to extricate you (i.e., your MPEC) from the brain-body in which you currently reside and implant it into a clone of you, then the clone would indeed be (a younger version of, but nonetheless) you.

However, if one were to bond your MPEC (i.e., you) to a clone of the brain and associated body within which another individual, for example, I (that is, my MPEC) currently resides, then that clone (of my brain-body pair) would be you and would not be me, by virtue of your MPEC's presence in the clone and my MPEC's absence from the clone in question. So, you—that is, at least for discussion's sake—let us hypothesize that you are your MPEC and you are not your brain-body pair (the one into which you were born).

Thus, you are not your brain or any other part(s) of the rest of your body. I am not my brain or body. I am my MPEC, MPEX MP or MP-CONEX. And any other person is that person's MP, not the brain-body, BB, in which it finds itself. And any brain-body pair your MPEX inhabits/ gets bonded to, becomes you. Any BBP or BB my MPEX gets implanted into becomes me. And any BB that any other specific one else's MPEC gets bound to, accordingly, becomes that specific (any) one.

DOES YOUR MPEC (MIND) AGE?

Understanding the aging process as, essentially, one whereby a living organism gradually falls apart into its (previously more or less effectively, structurally/functionally interactive) component parts, it would seem unlikely that a mind-particle-experiencer-consciousness (MPEC) would or could age, because, being a unitary, indivisible entity, an MPEC would have no component parts to fall apart into.

BY CIRCULATING EVERYONE'S MPEC THROUGH POSITIONS OF CLOSE PROXIMITY WITH EVERY OTHER MPEC AND LITERALLY THROUGH EVERY LIVING BBP, EVERYONE WOULD BE FAMOUS AND BETTER ABLE TO LIVE IN PEACE

By circulating (perhaps over and over again, unceasingly) every MP through positions of close (enough for mutual exploratory, edifying, interaction-permissive and -facilitative) proximity with respect to every other MP and also through (temporary and brief periods of bondedness within) every living BB on the planet, every MP would thereby come to know and appreciate the unique knowledge collection and special immutable, unique qualities of every other MP and the knowledge contained in and characteristics (though not necessarily immutable, by dint of always- and ever-possible genetic and other kinds of modifications) of or associated with every BB on Earth. Hence, everyone would be famous, because everyone, i.e., every MPEC, would be well informed and close-contactedly knowledgeable about every other MPEC as well as in relation to every living human BBP on the planet.

Consequently, worldwide fame would be one of the byproducts (even if not deemed a worthy goal in and of itself) that would accrue to every MP who would (want to and/or be wiiling to) partake of **global MPEC/BBP circulation**.

Due to the potentially (world-) widely-spread circulation not only of vast knowledge, but also, potentially, of each and every participant's feelings and deepest, heartfelt concerns, desires and aspirations, sincere cooperative motives leading to the formation of stronger than ever yet enjoyed bases for global harmony and peace might come into existence.

WOULD THERE BE ANY PRIVACY POSSIBLE FOR MPECs OPTING TO PARTICIPATE IN GLOBAL MPEC/BBP CIRCULATION?

Yes, there might still be the possibility of privacy. Each consciousness (MPEC), although a unitary entity, might nonetheless have internal differentiation into various structural and functional areas or inseparable parts, in much the same way as a statue representing a human figure, even if it is carved from a single piece of marble, might be differentiated into various structural regions or parts (e.g., head, arms, torso, legs, hands, feet, etc.), which respectively, perform the various, different-from-each-other, functional roles of representing and depicting the various and different parts of an actual human body.

For simplicity's sake, let us hypothetically visualize an MPEC, for example, as a unitary, hollow sphere, the topography of the internal and external surfaces of which may be differentiated into variously functioning, hill- and valley-like features, ridges, grooves, tunnels, helical structures, hammer-like appendages or extensions, stairway-like gradations, wheel-like or rotary areas, diaphragm-like regions, pulley, ratchet and conveyor-belt-like substructures, enclosures of various shapes, alternately expanding and contracting zones, etc. Let us suppose these diverse structural features, in their mutually cooperative and helpful roles, are capable of processing, sorting according to criteria of style and substance and storing (both within the MPEC itself and, possibly, also within the particular brain-body pair in which the MPEC under consideration is bonded) various kinds of experiences such as perceptions, information, memories, thoughts, emotional responses and other kinds of ostensibly subjective or experiential phenomena.

The kind of cubicles within the structure of each MPEC that might be used for (a) processing and storing encoded, perhaps encrypted, representations of highly personal, possibly appropriately maintained as private, educationally irrelevant, work-skills-irrelevant, generally intensely emotional experiences, perceptions, information, thoughts, memories, etc., might be expected to be structurally different from the kind of cubicles, also within the structure of each MPEC, used for (b) processing and storing encoded, but probably not encrypted, forms of relatively impersonal, perhaps businesslike, business-type or professional, mostly unemotional but educationally relevant and marketably valuable experiences, perceptions, information, thoughts, memories, concepts, skills and other objectively valuable phenomena.

Consequently, privacy could perhaps be preserved by and for all MPECs participating in MPEC/BBP (possibly worldwide) circulation, by dint of each individual consciousness (MPEC's) modifying its internal and external topographic features, shapes or structures, so as to create virtual, functional, locked lids or locked doors over the cubicles used to contain (a) -type phenomena (as denoted directly above), while leaving fully open and accessible (to and for all circulation-participant MPECs) all of the cubicles used to contain (b) -type phenomena (as also denoted directly above).

Similarly, each brain (part of each brain-body pair) might have (c) emotionally loaded, highly personal, educational- and employment-skills-unrelated phenomena, that is, the memory traces of such phenomena contained in functional, virtual or structural compartments of one kind which could effectively have locked lids or locked doors in relation to/from the standpoint of every MPEC (except the one originally born into the particular BBP). And (d) emotionally relatively neutral, educational- and employment-skills-related knowledge and information might be stored in another kind of functional, virtual or structural compartments that would be fully open and accessible to every MPEC passing through each particular BBP.

Hence, **privacy might remain as a durably reliable possibility**. And since each MPEC and each BBP might be expected to contain a unique collection of knowledge, memories and skills (with at least some of them being highly marketable or directly beneficial to oneself, i.e., to each one of us, individually), most (of us) MPECs might be expected to choose the quality-of-life (in particular, education and wealth-related) -improving effects that would be associated with worldwide MPEC/BBP circulation.

Nonetheless, it would seem reasonable, or even necessary, in view of notions of ethical rectitude, to acknowledge and facilitate each MPEC's right to make its own decision as to whether or not to (a) remain exclusively bonded to the brain-body into which it was originally born or (b) participate in worldwide MPEC/BBP circulation. If and when such circulation might prove feasible, then the marketable employment-relatable and directly self-benefitable knowledge and skills accessible to self-excluded, non-circulating MPECs would be so minuscule, compared to the quantities and qualities of these educational/vocational phenomena available to circulating MPECs, that non-circulators would, in effect, be relegating themselves to lives of (at best, relative) under-educatedness and poverty.

EVERY MPEC IS INHERENTLY, INTRINSICALLY AND IMMUTABLY UNIQUE

Some people (i.e., some MPECs) might worry that by their circulating through spatial positions of observationally facilitative, relatively close proximity with every other MPEC and circulating, literally, through every BBP, there would result a homogenizing, unique-identity-depriving effect on all circulation-participant MPECs.

For simplicity of discussion's sake, let us assume that each MPEC in the universe is intrinsically different from every other MPEC and unique, with respect to the entire universe, both subjectively, in terms of who it is and, correspondingly, objectively, in terms of what it is, from the standpoint of potentially observable characteristics. This assumption would seem to emerge readily from the realization that among all of the conscious entities (MPECs) in the universe, only one of them is you, i.e., your unique MPEC. Therefore, no appreciable reason would seem discernible as to how or why anyone's MPEC's uniqueness could be undermined in connection with its gathering any amount(s) or diversity/kind(s) of knowledge from any, many or all (other) MPECs and/or BBPs in the world or in the MP-BB-circulation-participant-system.

In summary, by means of planet-wide circulation of each (circulation-participating) MPEC through (positions of knowledge-and-information-sharingly-close proximity with respect to) every other participating MPEC and through every (MPEC-participant-corresponding) BBP, all of the participating (circulating), undividable consciousnesses (MPECs) and all of their associated brain-bodies (BBPs) would be known to each and every circulation-participating MPEC, hence, would be world famous, according as the number of participating MPECs might draw near, become close to or approximate the total global MPEC population number.

The following quoted observations (a–c) seem to support and strengthen the idea of a unitary, discrete particle of some kind of actual matter being the chemical/physical/medical-scientific basis of each person's unique, unitary mind (consciousness-experiencer).

(a) "The brain and body are built by DNA, and everyone's DNA is pretty much the same. We all have 99.9 percent the same DNA as Michael Jordan, Albert Einstein, Elizabeth Taylor, Charles Manson, Julius Caesar, Julia Child, and Jules Verne. All of them and everyone who has ever lived have the same 100,000 or so genes, which are organized into the same 23 chromosomes. But 'pretty much the same' is not exactly the same. There are differences in DNA—about 0.1 percent, or one bit out of every 1,000" (85). According to another source (86), "… there are thousands of tiny variations among individuals, but the overall variation is no more than 0.2 percent."

(b) "Human DNA is 98.4 percent identical with the DNA of chimpanzees and bonobos, a lesser-known chimpanzee-like ape. What is it in that other 1.6 percent that makes us different from them?" (87)

(c) "Which species is closer to chimpanzees: humans or gorillas? Obviously, chimpanzees and gorillas look very much alike. And humans look very different from both. So, naturally, everyone expected that chimpanzees and gorillas would be each other's closest relatives. But Sibley and Ahlquist took the two strands of DNA, zipped, heated … and found that chimpanzees were more closely related to humans than they were to gorillas. … For anyone who still doesn't like the idea that we humans are more closely related to chimpanzees than gorillas are, what can be done?" (88)

These observations would seem to tend to or might possibly support a mind particle theory of each person's mental identity, because it seems virtually inconceivable or implausible that all of the constitutional differences between any two randomly selected people or among the entire, apparently enormously variable human race or the much greater apparent differences between humans as a group and the "great" anthropoid apes (e.g., chimpanzees, bonobos and gorillas) as a group could be fully accounted for by genetic differences of only 0.1–0.2 or 1.6 percent, respectively.

Perhaps the seemingly unaccounted-for interpersonal (inter-human) variability and seemingly unaccounted-for inter-species (humans as compared with other, anthropoid primates) differences might possibly, potentially be accounted for, at least in part, by potentially observable and measurable high degrees of variability of characteristics or properties of MPEXes (mind-particle/consciousness-experiencers) between and among different individual humans and by major differences between the MPEXes of humans (as a group) and the MPEXes or MPEX-counterparts (if there are any) in the other species (as a group). And if the other anthropoid primate species' members do not each have any kind of unitary mind-particle/consciousness-experiencer, then the large human-as-compared-with-"great"-apes differences could possibly be, at least in part, accounted for in terms of the presence as compared with the absence of highly variable minds (MPEXes).

So, it is possible that from the moment when a (human) mind (an MPEX) gets implanted into a human fetal brain-body pair, BBP, to the moment of arrival at the point of full adulthood, the (perhaps highly inter-MPEX-variable) characteristics of this MPEX may exert or have exerted some powerful guiding influences on any, some or every aspect of the differentiation, growth, development and maturation of every cell in every human brain-body pair. These powerful MP-intrinsic guiding influences might also explain at least some or part of the reason(s) why genetically identical twins have biometric differences from each other, such as differences in their fingerprints, faces, and in the irises or irides (89) of their respective eyes, despite having identical genes (90).

A pair of conjoined twins (joined at the head), Laurie and Reeba Shappell, on whom a segment of a recent documentary television program (90) is based, seem to have two fully separate minds. They have significantly different personalities and do not directly or simultaneously experience the same or each other's emotions or feelings. For example, one twin said, in reference to the other, "When she has pain, I don't feel it. And when one of us is angry, the other one does not feel the anger; unless we're angry about the same thing." This is the case despite the brain-imaging-deduced fact that in many ways (neuroanatomically speaking) they share (apparently one, that is) the same brain. Or perhaps they have intricately-interwoven, but nonetheless (at least, according to brain-imaging equipment providing the highest currently possible resolution or power of visual discernment, under the twins' circumstances) extensively overlapping brain pathways, areas and volumes, especially in the frontal, temporal and parietal regions.

The (readily apparent, though, not necessarily genuine or true) incongruousness of two minds in, essentially, one brain might seem to further support and strengthen the idea that each person's mind (-particle-consciousness-experiencer) might be a separate and distinct chemical/physical entity. This entity or object might be bonded to and might spend most of its time traveling within the brain, but perhaps cannot accurately be equated with or conceived of as being the same object as the brain is.

ANOTHER WAY OF EVERY MPEC AND EVERY BBP BEING FAMOUS (AND PEACEFUL) WOULD BE VIA MPEC-INTERNET

Although MPEC circulation (the process of circulating each participating MPEC) through every living BBP and (through positions of close proximity to) every other (participating-in-circulation) MPEC might render all participating MPECs both highly knowledgeable as well as famous, a possibly less-energy-demanding (with respect to anticipatably high amounts of energy associated with rapidly and repeatedly making and breaking MPEC-to-MPEC-juxtapositional interactions and MPEC-to-BBP bonds of implantation) and a more energy-efficient method of virtual MPEC-BBP circulation, which might achieve the same results (gigantic knowledge and worldwide fame for all participating MPECs) might operate as follows.

Conceivably, every MPEC naturally forms a circuit with every (other) MPEC and also with every BBP on Earth, by means of (a) perhaps a virtually, infinitesimally slender outgoing (knowledge-seeking) appendage and (b) a similar, also virtually infinitesimally slender incoming (potentially knowledge-loaded and -conveying) string-like or thread-like process, branch, arm-like appendage or interconnection, with (a) and (b) together constituting a loop or circuit between every two MPECs and between every MPEC and every BBP.

If this is true, then the reason why we do not seem to perceive these interconnections might be inferable from the explanation about to be presented. Each MPEC-to-MPEC, MPEC-to-BBP and BBP-to-MPEC (almost infinitesimally slender) interconnection might contain a mechanical (on-off, close-open) switch. Most of these interconnections currently are (and in the recorded history of our planet apparently always have been) firmly fixed into the information-no-flow, off or open circuit condition. Hence, knowledge, pleasure (and displeasure) are not automatically shared among all MPECs. On the contrary, when any one of us experiences, feels or appreciates any of these phenomena, we generally, very often or usually perceive ourselves as experiencing them in an individuated if not in a solitary way, relatively disjoint from the knowledge, pleasure (and displeasure) of others.

Nevertheless, according as the switches might infrequently, intermittently and spatially-sparsely close into an information-pro-flow, ON (as opposed to OFF), or closed-circuit condition or configuration, then there might be, by this means, a genuine structural, biomedical, physiological basis for the **Collective Unconscious**, perhaps as postulated by Carl Jung (91, 92). This sporadic closed-circuit condition may also be the basis of the hypothesized interconnectedness of all particles in the universe, as suggested by experimental findings (93–95). Relatively infrequently occurring and, sometimes, brief instances of these switches being closed and, therefore, permissive of all kinds of informational and emotional inter-flow might be at least the initial bases or catalysts of love, friendship, mutual empathy and intuitive knowledge.

However, considering the possibility that such MPEC-to-MPEC, MPEC-to BBP and BBP-to-MPEC interconnections might not naturally exist, a question arises: Is it conceivable that it might behoove us to muster and implement the motivational wherewithal and technological expertise (culled from within the altruistic and mechanically-inclined resources of our wisdom and knowledge) that would be necessary to construct and activate such interconnections among all of us who might be willing and wanting to live in highly harmonious, mutually informative and educational, sharing and caring ways? The answer might conceivably, sooner or later, and arguably auspiciously, be yes!

It would seem reasonable to infer this affirmative answer regardless of whether the universal-harmony-conducive task be one of (a) discovering MPECs and (subsequently discovering) naturally existing, interconnecting one-MPEC-to-another-MPEC-then-back-to-the-first-MPEC and MPEC-to-BBP-then-back-to-the-same-MPEC circuits and then determining and implementing means whereby to close the circuits and thereby actuate and sustain flow of harmony-facilitative phenomena or a task of (b) discovering MPECs and subsequently constructing—out of some, possibly, human-made material or other kinds of interlinking substance(s)—the requisite interconnections which would be designed and implemented in such ways as to permit many kinds of harmonizing inter-flow. In either case (a or b), the prospects of inducing numerous kinds of constructive developments might be reassuringly boundless! Let us refer to a/any one-MPEC-to-another-MPEC-then-back-to-the-first-MPEC circuit as an MPEC-through-MPEC circuit. And let us refer to a/any MPEC-to-BBP-then-back-to-the-same-MPEC circuit as an MPEC-through-BBP circuit.

If it were possible to manufacture and put in place virtually infinitesimally slender, ON/OFF-switch-containing interconnections (a) between each MPEC and every other MPEC, (b) between every MPEC and every BBP and (c) between every BBP and every MPEC, then by the simple action of opening and closing the ON/OFF switches contained in the interconnections, it would be possible to achieve virtual, as opposed to actual, circulation which would be analogous to the Internet of computer-associated renown.

Let us suppose the world's human population were ten billion people (i.e., ten billion MPECs, with each of us MPECs bound within our own corresponding BBP into which we were, respectively, born). Then, assuming all of them (us/we) were to want to participate in the MPEC-BBP-internet, that is, participate in virtual MPEC-BBP circulation, then each of us, that is, each of our MPECs would have twenty billion connections or, synonymously, segments (with each full circuit being composed of two segments, one for (a) seeking and locating information/knowledge and another for (b) receiving, securing and consolidating information/knowledge, thereby, together (a and b) constituting a complete, functionally or effectively (loosely speaking) circular pathway for each MPEC-through-MPEC circuit) to and from the other ten billion (actually, ten billion-negligibly-minus one, with this one representing an unnecessary circuit with itself) MPECs and twenty (additional) billion connections (or segments) to and from the ten billion living human BBPs.

So, the total number of interconnections or, equivalently, segments would be:

10^{10} x (2×10^{10}) + 10^{10} x (2×10^{10}) = (2×10^{20}) + (2×10^{20}) = 4×10^{20} interconnections or segments, with each containing an ON/OFF or CLOSED-/OPEN-circuit switch. Hence, there would be half of this number of complete two-interconnection- or two-segment-containing circuits or 2×10^{20} complete circuits, within or among the participant population as a whole, with this number comprising 10^{20} MPEC-through-MPEC circuits and 10^{20} MPEC-through-BBP circuits.

During any finite duration or interval of time (the specific length of which might be determined as the resulting average value of a one-vote-apiece- or one-vote-per-participating-MPEC selection process, but for illustrative purposes, let us consider for example, during each full one second of time [i.e., one-sixtieth of a minute]), each MPEC might have each one of its ten billion MPEC-through-MPEC circuits and each one of its ten billion MPEC-through-BBP circuits sequentially turned on and then off, in series, i.e., only one-ON-at-a-time (that is, switched to the closed position) once, for a turned-on duration of one twenty-billionth of a second (and a turned-off duration of one second minus this one twenty-billionth of a second) for each one of every MPEC's twenty billion circuits. Hence, during each twenty-billionth of a second of time, each MPEC would be gaining knowledge and other information from and, simultaneously, imparting knowledge and information to another entity or object (i.e., one of the other conscious MPEC[s] or the knowledge/information storage facilities-compartments or other kind[s] of receptacles of one of the BBP[s]).

And within each one-second-long interval, each of the ten billion MPECs (i.e., each of us human minds) would have gained knowledge and other information from and imparted these valuable phenomena to all of the other ten billion (minus one) MPECs and all of the ten billion BBPs. Such a system would constitute an MPEC-BBP-internet or, synonymously, a process of virtual MPEC-BBP circulation (abbreviatedly designatable as MPEC-internet or virtual MPEC circulation) that might be expected to entail less necessary consumption or expenditure of energy than actual MPEC circulation, while sharing (obtaining from and imparting to BBPs and other MPECs, i.e., MPECs other than oneself) the same amount of knowledge and information as in the case of actual MPEC circulation.

This kind of virtual MPEC circulation (or MPEC-internet) might appropriately, alternatively, be dubbed (MP-CONEX-, MPEX-, MPEC- or) MP-internetus, MPEX-internet-us, MPEC-inter-connect-us, MPEC-inter-connectus, MPEC-interjoinus, MPEC-interjoynus, MPEC-interlinkus, MPEC-internetus, MPEC-internexus, MPEX-interconnectus, simply, MP-interconnectus, etc. Analagously, actual MP circulation might be referred to as: MP-intercirculatus, MP-CONEX-intercirculate-us, MP-inter-perfuse-us, etc.

HOW EVERYONE MIGHT BE PAIN FREE

The seemingly widely-held view that pain is as necessary a part of conscious experience as pleasure is, may prove incorrect. The question, heard more than once by this author, "How would you know pleasure if you did not know pain?" seems only as rational (which is to say, not very rational) as the question, "How would you know love if you didn't know hate?"

All of the adaptive value of pain could possibly be achieved with, and functionally replaced by, gradations of intensity (and various qualities or kinds) of pleasure. For example, suppose a person's body becomes diseased in some way. The disease (currently) produces pain. At times, this pain might become so intense as to be intolerable and, therefore, constitute uncontrollable pain: i.e., suffering, that is, pain so intense that the afflicted individual cannot control impulses leading toward suicidal ideation. The diseased person, in response to the pain, visits a physician, or other health-care professional/practitioner, who prescribes medication which cures or effectively treats the disease and fully relieves the pain (suffering and suicidal ideation).

However, the same fortunate result could conceivably occur without pain (or suffering) necessarily being involved, as follows. A person's body becomes diseased. The disease produces no pain (or suffering), but does put a stop to any experience of pleasure. In this case, the diseased person, alerted by and in response to the cessation or, at least, marked diminution of pleasure, might be expected to visit a physician or other treatment provider who might prescribe a medication that might cure or effectively treat the illness, and, consequently, restore the patient's pleasure and enjoyment of life to their usual, amply high levels.

So the same curative or adaptively treatment-effective results might be expected to occur in connection with varying, but completely painless (and completely without and free of suffering) intensities or degrees of absence or presence of pleasure, which could serve as varying intensities of signals for proportionately (to a relative absence or inversely proportionally to a relative presence of spontaneously/naturally occurring, good-health-associated pleasure) appropriate degrees of alertness, sustained vigilance and medical intervention, just as might occur with a hedonic spectrum including pain or actual discomfort of some/any kind(s).

While serving as an analogy in relation to the not-necessarily-valid notion of pleasure being appreciable only through the contrasting experience of pain (or suffering), the following example may also serve to rebut any notion of love being appreciable only through a contrasting experience of hate. Suppose person A receives an invitation to a party from person B (whom A hates) and another invitation from person C (whom A loves). Understandably, A will probably decline B's invitation and accept C's invitation. The very same result (acceptance of C's invitation and rejection of B's) could be expected if A were to love C and feel merely neutral (but not at all hateful) toward B.

These examples, hopefully, may be convincing of the ideas that pain is not a prerequisite for the appreciation of the goodness (enjoyableness) of pleasure or a prerequisite for an individual (an MPEC) to function adaptively (with respect to danger, risk or threat). And hatred is not a prerequisite for the goodness of love to be appreciated or for an individual to function appropriately in relation to both loved and not-especially-loved other individuals.

WE MIGHT CURRENTLY OR EVENTUALLY RENDER OURSELVES INCAPABLE OF EXPERIENCING PAIN OR SUFFERING BY MEANS OF BSNP OR (INSTANTANEOUS) GENETIC SELF-REENGINEERING (IGSRE)

For example, by extrapolating suggestive if not definitive evidence (20, 96–102), it might be possible to validly infer that **(BSNP) brain stimulation/neural pacemakers/prostheses** might be implemented in such ways as to pervasively eliminate virtually all kinds of pain and suffering (with "suffering" being defined, for this context, as uncontrollably intense pain). This goal might be accomplished by means of (a) surgically implanted electrodes or, preferably, (b) surgically noninvasive [e.g., electromagnetic, ultrasonic, waveform-mediated or other kinds of chemical-physical modalities or means, perhaps targeted-to-specific-neuroanatomic-sites/receptors, etc.]. Of course, we would want to include in our repertoires of pain/suffering-alleviative-BSNP, effective devices/mechanisms whereby to bring about patterns of pleasure-intensity-variability that might enable us to take appropriate fight-or-flight action whenever the continuity of our pleasure might be threatened or endangered, as explained above.

Alternatively to BSNP (or used together with it, in ways that might increase the overall therapeutic, pain/suffering-alleviative efficacy), one or more genetic engineering approaches, such as the one suggested presently, might be effectively implemented.

By extrapolating the genetic common denominators (of individuals with high thresholds of pain and suffering of all kinds) upwards into ranges where no perceptions of pain or suffering are possible and then inserting the necessary modifications into the genes and chromosomes of all the cells in our brain-bodies (103), we might instantaneously render ourselves incapable of experiencing any kind of pain or suffering. Of course, we would also want to insert, into our genotypes, effective programs of pleasure-intensity-variability, which would enable us to take appropriate protective fight-or-flight action whenever the continuity of our pleasure might be threatened or endangered. The notions of using brain stimulation/neural prostheses and/or of genetically engineering ourselves in such ways as to preclude any experience/-ing of pain or suffering are explored in some detail in an earlier paper (104).

A BSNP-based approach, by which the entire human race might readily and markedly benefit in terms of relief of pain and suffering, seems to be currently feasible (1–70, 96–102), as might also be true in the case of learning- and work-skills-facilitative (LWF-)BSNP. In view of existing knowledge and already-accomplished research and development (1–70, 96–102), it seems probable that both pain/suffering-alleviative and learning/work-skills facilitative BSNP could be brought about in a mere few or several years' time, at a cost that would probably be in the multimillion dollar (but not the billion-dollar) range (references: the same as directly above). Alternatively, genetic-engineering or gene(tic)-therapy-based approaches to pain/suffering relief and learning- and work-skills facilitation (103) seem to be potentially (or, perhaps, even only eventually) but probably not currently readily altogether feasible, at least not in immediately or currently producible, dramatically pain-and-poverty-alleviating ways.

HOW WE MIGHT BECOME DEATH FREE OR IMMORTAL

In view of each of our MPEXes possibly being ageless, as hypothesized above, one way in which we could conceivably become deathless or immortal would be by transferring to a new clone of the original brain-body into which we, respectively, were born, whenever the most recently issued BBP-clone might become too aged or otherwise unfit to continue living.

Another way of freeing ourselves from the prospect of death, one that would be educationally and informationally far more interesting than the way noted in the directly preceding paragraph, would be by means of MPEX (mind-particle or experiencer) circulation or virtual mind particle circulation (MPEX-internet) as explained above. Nonetheless, even with MP circulation or MP-internet, too-aged-to-continue-to-function or otherwise unfit-for-continuing-to-live BBPs would need to be replaced by cloning, genetic engineering, (neural-) stem-cell-entailing or some other kind(s) of reproductive or regenerative methods (105–110). Via MPEC circulation or internet, one would be continually and unendingly gathering the educational and informational benefits of billions of different MPEXes and billions of different BBPs, instead of being limited to the relatively meager range of possibilities that would be associated with the same BBs being cloned over and over again.

HOW EVERYONE MIGHT BE VERY WELL EDUCATED

Relative to current educational levels and occupational standards, substantial augmentation and enhancement of individual per-person (i.e., per-MP) knowledge accumulations and work-skills-related repertoires might be brought about or catalyzed by methods that could be (a) probably de-effortized, (b) possibly pleasure enjoyitized, and (c) conceivably, potentially, engendered in such ways as to occur seemingly automatically, instantaneously and below the threshold of conscious awareness. These methods would preferably consist in or entail (at least, macroscopically-speaking) surgically-noninvasive, deep or superficial brain-/neural-process-facilitative, computerized, (preferably micro-miniaturized, nanotechnologized) neural-function-modifying or -modulating implants or nonimplanted/non-implantation-requiring devices, mechanisms, modalities, or means.

They might all be considered appropriately placed within a category of BSNP (brain stimulation/neural prostheses) together with (or perhaps without) pharmacologic performance-improving agents and/or pharmacotherapeutic/medicinal substances. These methods/treatments might be, collectively, mutually-potentiatingly/synergistically helpful, simply additively-beneficial, or perhaps even singly, individually-adequately effective, so as to qualify most of us as very well educated.

Further improvement of educational and vocational potential and achievement might be brought about by methods whereby any consciousness or conscious entity (MPEC) bonded to any brain-body pair, BB, could (at any point[s] in time during its period of occupancy, inhabitation or bondedness within that particular BB, whether that period be 100 years, one twenty-billionth of a second or any other duration) very rapidly or instantaneously (as direct consequences of the conscious entity's, i.e., the MP's making selections and implementations entirely in accordance with its own free will and choice) either add (insert into the BBP's genotype) or delete (eject from the genotype) and virtually-simultaneously activate or deactivate genes of any quality-of-performance-related or of any other quality-of-structure-or-function-related kinds(s) (103).

Hence, any consciousness (i.e., MP or MP-CONEX) bonded within any brain-body pair could at any chronological point(s) during and throughout its period of bondedness therein, in accordance with its exercise of its own intrinsic free will, improve (according to its own criteria of what constitutes improvement) the effective intelligence, educational and vocational wherewithal of the BB in question by instantaneously changing any, many or even all of its (the BB's) characteristics by instantaneously making genetic alterations within any, many or all of the BB's cells. This freedom and capability would constitute instantaneously implementable genetic self-reengineering. It would perhaps be accomplishable via (a) gene pills, possibly entailing the use of modified, specially adapted bacteria, viruses, transposons (76), plasmids (73, 76), etc., (b) aimed-to-precisely-predetermined-location particle beams or (c) other means.

This kind of spontaneous, ongoing, never-immutable genetic engineering might be a substantial boon to the prospect of self-improvement, as part of an overall panorama of pursuit and possibility, and especially a boon to the prospect of educational/vocational self-actualization and fulfillment. Ramifications of this boon might include all aspects of the state of being and process of doing, or becoming something and someone better than ever before, according to one's own standards and criteria, including even aspects pertaining to purely cosmetic properties, physical appearances of any BBP and related phenomena such as height, weight, sexual attractiveness, as might be associated with the nature or degree of development of primary or secondary sex characteristics, etc.

HOW EVERYONE MIGHT BE *EXTREMELY* WELL EDUCATED

In order for us to better keep pace with computers, which apparently are becoming "smarter by the minute" (111), it might behoove us to consider option A: perpetually-recycling MPEC circulation or MPEC-internet (virtual MPEC circulation) of each and every one of us (MPECs) who choose to participate in being repeatedly cycled through, that is, throughout the full gamut of all of the ever-changing, ever-increasing knowledge and information accumulations, aggregates or collections contained within each of the billions of individual mind particles (MPECs) and each of the individual brain-body pairs, BBPs, because option A might serve us better than option B: brain stimulation/neural prostheses, pacemaking, even when/if used together with instantaneously effectible genetic self-reengineering.

However, it may be reasonable to expect that the relatively biggest and best advantages might be enjoyed by combining the use of both options A and B. Additional educational enhancement and knowledge-augmentation within each (consciousness, mind or experiencer, that is) MPEC might be brought about/facilitated by repeatedly cycling/circulating or interconnecting each (circulation/MP-internet-virtual-circulation-participating) MPEC or MP not only through the knowledge collections/aggregates contained in BBPs and those contained in other MPs, but also through (the data, especially potential or actual employment-relatably valuable material/information that might be contained in enormous quantities within the fundamental underpinnings of the information-explosion, in particular, within and available by means of traversing through) computers themselves.

If we were to pose the question, "How highly or how well educated might all of us (MP circulators or MP virtual circulators, that is, mind-particle-internet participants) be if the MP circulation cycle or MP-internet were to merge a portion of itself, that is, a portion of its cycle were to be spent within the information-loaded realms of actual computers and especially within the realms of the actual computer Internet / global information superhighway / World Wide Web?" A possibly valid answer might be: "Stratospherically highly, encyclopedically or comprehensively well educated."

HOW EVERYONE MIGHT BE SEXUALLY LIBERATED

Brain stimulation/neural (possibly pacesetter or pacemaker) prosthetics (BSNP), whether involving excitatory or inhibitory neuroprostheses, depending on the choice of the individual, might be helpful, as follows. Its value might be appreciable when applied to sexual-impulse-subserving/-mediating pathways in the brain, while using appropriate, effective stimulus parameter values. BSNP might be used to enhance erotic or sexual impulses one wishes and chooses to feel and act on (with enhancement designatable as a libiditron mode of BSNP therapy), possibly even to the point(s) of full gratification or orgasm(s), single or multiple in occurrence, with an orgasmic mode designatable as an orgasmatron (111, 112) mode of BSNP. Or it might be used to suppress sexual impulses one would prefer not to be compelled by, with appropriate, adequately effective inhibition being designatable as a sublimitron mode of BSNP therapy. This mode of application might be helpful whenever one feels sexual behavior would be inappropriate to context or whenever one would prefer to divert or sublimate arousal into some nonsexual cognitive process and its associated behavioral expression.

Instantaneous genetic self-reengineering might add on another more powerful dimension of capability in relation to erotic/sexual self-determinism, again, in accordance with individual choice and MPEC-intrinsic free will.

HOW EVERYONE MIGHT BE HEALTHY

The augmentation and enhancement of global medical knowledge, as might be anticipated to result from BSNP, instantaneously-effectible-genetic-self-reengineering (IEGSRE), MPEC-BBP circulation and MPEC-BBP-internet, especially when and if MP-circulation/internet might be interfaced or internetted with a global computer Internet, its successors or other future derivative developments, could possibly entail advances in medical treatment that would be substantial and comprehensive enough to encompass everyone comfortably within the category of being thoroughly healthy. Additional and more instantaneously impressive universal-good-health-promoting-and-maintaining phenomena and, in particular, processes might be entailed in ways implicit as follows.

The aforementioned education-augmenting and -enhancing modalities might accelerate and magnify the trend toward development of micro-, mini-technology, describable as nanotechnologization, or proliferation of nanotechnologies, to such an extent as to engender conception and implementation of microscopes and imaging modalities so much more powerful than any currently available that we might actually be able to literally see (or otherwise perceive) molecules, atoms and even subatomic particles. Consequently, we might be enabled to literally see and otherwise perceive individual minds (i.e., conscious experiencers or experiencing entities, conceptualized herein as mind-particle-consciousness-experiencers, MP-CONEXES). Then, we would be seeing what we really look like, that is, how our respective essences of individual identity (MPs) actually appear and what each and everyone's objectively viewed, measurable, quantifiable boundaries of self can accurately be delineated as. It would be astonishing, but not altogether inconceivable, that we (our MPECs) might be discerned as manifesting some significant degree(s) of resemblance to the macroscopic people we, respectively, observe in our mirrors.

With vastly increased powers of observational magnification accessible to us, we might be able to literally see or otherwise perceive (at each and every three-dimensionally specified and located point in the human body) exactly how, that is, precisely what are the mechanisms whereby molecules, atoms, subatomic particles and waveforms interact with each other in such ways as to constitute or mediate diseases and disordered conditions and dysfunctional processes.

Consequently, we might be able to readily deduce, directly from our perceptual observations, how we might most beneficially, even curatively design, create and deliver medications and other therapeutic interventions and modalities, rather than formulating medical treatments based on relatively indirect, often tenuous, experimental and clinical inference and reasoning, as is necessary when crucial mechanisms of disease cannot be actually, directly observed or perceived.

HOW EVERYONE MIGHT BE ANOREXIA-, OBESITY-, INSOMNIA-FREE, ETC.

By means of selective activation and/or deactivation, i.e., excitation and/or inhibition of appropriate site(s), pathway(s) or structure(s) in the brain or elsewhere in the nervous system, brain stimulation/neural (possibly, pacemaking or pacesetting) prostheses (BSNP) might give rise to and enable relatively de-effortized and reliable dieting, which would readily empower anyone to easily, reliably and predictably lose or gain as much weight as desired. There might be no reason or occasion to experience the stressful, strenuous hardship and unreliability of needing to generate and sustain will power or to experience dietary deprivation, hunger or food craving to any degree or in any way whatsoever.

Effort-free weight control, modification and maintenance might be attained to / achieved by stimulatively (activatingly or deactivatingly) focusing on the brain's ostensible (a) feeding/dietary-appetization-subserving and/or (b) dietary-satiation-subserving regions or centers, such as may, respectively, be located in the (a) lateral and (b) ventromedial hypothalamic nuclei (113).

Additionally, if one were displeased with one's (own) body shape, contours or body habitus, as associated with some particular pattern or distribution of body fat, then, perhaps without there being any need for or appropriateness of changing dietary intake, one might be able to use (surgically) noninvasive focused electromagnetic, acoustic/sonic, thermal, particulate (consisting in one or more particle[s] of some type[s]), or some other kind(s) of stimuli that would be directable to volumes or pockets of relatively large accumulations of fat in such way(s) as to (1) selectively increase the rate of effective, local fat metabolism or (2) decrease the efficiency of localized absorption of fat-storing nutrients, either way (1 or 2) so as to cause attrition of fat-storage/

deposition only within the narrowly or selectively circumscribed body areas where one were to feel one might have too much adipose tissue. Effective techniques might be capable of expeditiously, maybe even instantaneously, dissolving, melting or otherwise reducing and re-contouring the self-determined, body-topographically-relatively-localized or distributively disproportional, excess adiposity, without affecting those areas of the body with respect to which one might feel the amount of fat deposition is aesthetically optimal or, at least, acceptable to oneself.

Conversely, if someone were to feel there were some part(s) of one's (own) body wherein there were not large enough accumulations of fat (for example, in too-thin legs), one might be able, here, also, without any change in dietary intake being necessary, to direct preferably (surgically) noninvasive stimuli that would be modified, in terms of those aspects mentioned in the preceding paragraph, in such ways as to (a) decrease the rate of localized fat metabolism or (b) increase the efficiency of localized absorption of fat-storage-deposit-augmenting nutrients, either way, or in either case (a or b), so as to increase the amount of adipose deposition and bring about the associated, desired change(s) in body habitus.

BSNP might also facilitate full pleasurization and defortation (de-effortization) of all kinds of bodily exercising and bodybuilding. Hence, we, most of us anyway, might have more muscular and generally better-maintained, better-toned bodies. An additional, perhaps BSNP-mediated, efficiency-of-time-utilization-improving benefit might be that a person would possibly be able to simultaneously (a) automatically, unconsciously or subconsciously, virtually entirely defortizedly (effortlessly) exercise, or perhaps be passively exercised, perhaps by means of functional electrical stimulation, FES (61), of one's body musculature and other body components, such as pulmonary, cardiac and vascular aspects while (b) the person's brain and mind (i.e., mind-particle-experiencer-consciousness, designatable as MP-CONEX, MPEC, MPEX or MP) might be actively, fully-consciously engaged in valuable, constructive and readily marketable learning and working. Also of possibly valid relevance to the overall area of concern/pursuit in respect of well-developed body musculature may be the notion of being able to increase and improve body-muscle mass, strength and condition via specially developed drugs capable of increasing muscular development perhaps even in the absence of, that is, without any exercise being necessary (114).

Perhaps by means of BSNP-mediated focusing on, exciting or suppressing of appropriate brain or neural site(s), regions or pathways, it might be readily, easily, painlessly, and virtually instantaneously possible for anyone to stop being addicted to any substance(s) without having to invoke or sustain difficult-to-maintain self-denial, resistance to ceaseless temptation, and without having to endure (1) the physical-dependence-associated symptoms of withdrawal, (2) tolerance (the need to progressively increase the dosage or intake of many or any medicinal or other substance(s) in order to achieve the same remedial effect), (3) the mental-dependence-associated symptoms of craving the addictive substance or (4) even the feeling of being depressed by reason of being deprived of something strongly desired.

BSNP might be used to alleviate motion sickness and nausea (in general) whenever it might occur, such as can be mediated by the vestibular apparatus of the internal or inner ear. For example, nausea might be relieved by means of inhibitory stimulation or deactivation of the "vomiting center" located within the medulla region of the brain (113, 115, 116) known as the area postrema, where there is no appreciable blood-brain barrier (116), to act as an obstacle to medication or any other therapeutic agents or modalities.

And BSNP might also eliminate or markedly diminish the hurtful aspects of receiving criticism, so that we would possibly be better able to benefit from it, without being traumatized in any way(s). Moreover, BSNP might help the blind to see (39, 43, 117–119) and the deaf to hear (39, 43, 113, 118, 120). BSNP might be therapeutically valuable for anyone who has been deprived of the use of any of a number of various bodily functions, for example, by dint of stroke(s) (61, 73, 113, 119–122) or injuries (123, 124). It might significantly help them with their lost or diminished faculties or functions by activating or deactivating functionally appropriate and useful site(s), structure(s), or pathway(s) in the brain, elsewhere in the nervous system or even outside of the nervous system. The prospects of BSNP-mediated or BSNP-facilitated compensatory or alleviative treatments, if used together with other methodologies that might bring about neurogenesis, nerve regeneration, other kinds of cellular, histologic or whole-organ replacement/redevelopment, perhaps via pluripotent, neural or other types of stem cells, tissue transplants (105–109, 121), etc., might be applicable in such ways so as to be observed to be of notable value to neurologically deprived or otherwise medically challenged/impaired individuals.

BSNP medical devices might contain variable, rheostat-like and ON-OFF-switch-alarm-clock-like, modulating mechanisms, effective with respect to alertness-vigilance versus sedation-relaxation levels, so as to provide the following benefits. In particular, one valuable possibility might be any person's instantaneously implementable, self-willed, self-controlled, merely BSNP-mediated induction and selected modes of maintenance of deep, neurophysiologically more-than-adequate, even replete sleep. The sleep-inducing and -regulating mechanisms might be expected to be thoroughly dependable.

Accordingly, anyone could obtain such high-quality sleep virtually anytime, anywhere, for precisely as long (and no longer a period of time) as the individual might need or want it, in order to attain full refreshment and reinvigoration. And this BSNP approach could be feasible despite any and all circumstances and concerns as are conceivable as being currently capable of unduly and maladaptively interfering with and undermining a person's quantity or quality of sleep and degree of restedness.

Sleep-inducing BSNP, perhaps entailing focused, targeted substances/chemicals/particles (pharmacologic/medicinal agents) and/or electromagnetic, acoustic, etc., focused fields, waveforms, beams, streams or, perhaps, tidy "lines" of precisely aimed molecules, atoms, ions, subatomic particles, etc. (125) might conceivably involve activation of a small cluster of cells in the ventrolateral preoptic (VLPO) area of the brain's hypothalamus (126), which has been referred to as a "pinpoint (ON/OFF) sleep switch." These cells, when they become active, send inhibitory signals to the brain's arousal/alertness-producing/-subserving system and thereby bring about sleep. Other approaches to reliably producing health-sustaining/improving sleep might entail relatively diffuse stimulation of neuroanatomic sites functionally entailed in spontaneous, unassisted sleep and stimulation, conceivably, of some neuroanatomic sites used in general anesthesia.

Moreover, most kinds of physical/bodily pain and suffering, irrespective of pathogenesis (for example, in association with "advanced carcinoma," as per page 225 of ref. 20) and irrespective of the nature of specific symptoms that have been observed and reported (20, 23, 96–102), have been found to be substantially remediable with BSNP. Mental pain and suffering also seem to be encouragingly responsive to this mode of treatment (4, 19, 20, 22, 23).

Noteworthy prospects as potential components or aspects of processes that might prove useful in conceptualizing and applying systems of BSNP might include adaptations of electromagnetic, acoustic, biochemical and pharmacologic media, modalities and approaches, as might be used in such ways as to provide effective, highly focused (or, as appropriate, relatively diffuse) treatments or performance/outcome-enhancers (125) for various medical conditions, diseases, disorders, syndromes and pathogenic mechanisms involving, occurring or functioning in any neural or nonneural anatomic regions, body tissues, structures or organ(s') system(s).

Hence, such prospects might potentially constitute categories of therapeutic formulations or treatment methodologies that might be designated as function-modifying pharmacologic, electronic or otherwise characterizable means of neuronal, neural, or nonneural modulation.

Micro-miniaturized therapeutic or otherwise beneficial (such as performance-improving) robots, nano-robots (42) or, referring to a term used by Raymond Kurzweil (43), "nanobots" might be conveyable and deployable as strategically useful, neuroanatomically or otherwise anatomically precisely positioned entities. Passageways of body-entry might include those of inhalation, injection, permeation of inter-molecularly, even inter-atomically-sharply-focused penetration/propulsion (125) that is, passageways that would be targeted and utilized for good-health-conducive or bodily-function-improving substances (e.g., particles) or waveforms. Via BSNP, instantaneously effectible, genetic self-reengineering (IEGSRE) and MPEC-BBP-circulation-internet, virtually all diseases, disorders and other afflictions might be anticipated to be viably escapable (that is, any afflicted individual MP-CONEX/MP/MPEC/MPEX might be able to deathlessly or undyingly escape from the afflicted BB/BBP and transfer to a healthy, unafflicted BBP) even if the affliction itself (of the original BB) were not readily, effectively treatable or curable.

TO BE WHATEVER WE WANT TO BE, WHENEVER WE WANT TO BE

By means of the methodologies included in the preceding paragraph, each and every one of us (each mind-particle-experiencer-consciousness, MPEX, or mind-particle-consciousness-experiencer, MP-CONEX) might be able to inhabit or, synonymously, be bonded to brains and bodies that fit whatever descriptions we want them to fit, whenever we want them to apply. We might be able to change any mental or physical characteristic(s) we want to change, perhaps instantaneously, via precisely targeted waveforms and/or substances, in almost any way and to almost any degree we might possibly desire (104).

Hence, each of us (MPEXes) might be able to (possibly, instantaneously) morph (i.e., change the characteristics of the BBP we find ourselves, respectively, ensconced in, that is, change into any form or change any aspect[s] of the overall morphology, structural minutiae and/or physiological functioning of) any (or even all) of the attributes, internal components, clinically, cognitively, or adaptively significant processes or external parts of any brain-body pair (perhaps by genetic splicing, insertion, deletion, activation, deactivation or, perhaps, by other kinds of techniques) at any time during any portion or throughout any period of our (MP's) occupancy of, being bonded or bound to, or implanted within any brain-body in question or under consideration.

CAN COMPUTERS BE CONSCIOUS?

Let us suppose a mind or (in other, but equivalent, words, a mind-particle-consciousness-experiencer, MPEC, MPEX or, simply, abbreviatedly) an MP can be bound, bonded to or implanted within a computer. In particular, let us suppose an MP can be attached therein in such a way as to be able to follow the direction of flow of information within and keep pace with such rapid flow of information, as generally occurs within a modern computer. Then, the answer to the question of whether or not computers can be rendered conscious would apparently be revealed to be "yes," computers can (perhaps by being implanted with an MP) become conscious. Then you, I or anyone else might be able to find out directly (by firsthand experience) what it would be like, what it would feel like to actually lead the life of a (conscious) computer. Such an opportunity might occur as a consequence of your, my or anyone else's (MP's) allowing and facilitating the implantation of our respective mind (that is, consciousness or MP) into the confines of some kind(s) of chemical, physical bond(s) within (at least, some components of) a computer, for some appreciable period of time, perhaps hours, weeks or years. Then the conscious computer might be regarded, at least by some people, as a new life-form on Earth.

Being disembodied and separated from its human brain-body, it would probably seem incongruous, at least to some people, to regard a computer-implanted human mind (MPEX) as a veritable human being. Yet it would probably be relatively easy to regard an MPEX-containing computer as being alive (hence, some kind of form of life), by reason of its experiencing (various phenomena, collectively, constituting) a life, even though neither an animal's nor a plant's nor any other conceivable kind of organism's biologically-based type of life. So, the answer to the question of whether or not computers can be or can become conscious, perhaps as reasoned directly above, is possibly "yes."

HOW WE MIGHT ALL BECOME RELATIVELY UNSELFISH

Anyone who is desirous of becoming relatively more unselfish might be able to do so by using BSNP to enhance compassion for and empathy with others (26). Voluntary genetic self-reengineering (perhaps of an instantaneously effectible kind) might be expected to add another dimension of capability to the possibility of attainment to a goal of relatively increased unselfishness. MPEX-BBP circulation and virtual circulation (MP-internet) by their very mutually (interactive) nature would almost unavoidably enable or induce all participating MPECs to become more empathetic, sympathetic and unselfish than we currently are.

These increases in unselfishness might occur as readily understandable consequences of, and in direct connection with all MPEC-BB circulation-internet (that is, MPEX-BBP circulation and/or virtual circulation, MP-internet) participants' cycling through the entire population of participating MPs, BBs and computers included in the MP-BB circulation/internet loop, over and over again, and thereby directly experiencing the needs and wants of MPECs in addition to our respective self, and the strengths, weaknesses and problems of all of the BBs (in addition to the one each of us was, respectively, born into) and of all of the loop-included computers, far more convincingly-vividly and tangibly than is possible with each of us MPECs being confined and biochemically-biophysically-medical-scientifically bonded (in some yet-to-be discovered way) within the single, individual brain-body pair, BB, in which we, each of us, respectively, generally seem to find ourselves quite thoroughly contained.

HOW WE MIGHT PROCEED TO RESEARCH AND DEVELOP BRAIN STIMULATION / NEURAL PACEMAKING, NEUROPROSTHESES, MIND-PARTICLE-EXPERIENCER-CONSCIOUSNESS DETECTION, ISOLATION AND CIRCULATION/INTERNET

Electrical/electronic brain stimulators, neural pacemakers/pacesetters or stimulating electrodes/microchips have been and are currently in effective use as surgically implanted devices, as well as functioning, in some cases, as completely surgically noninvasive devices/equipment of various types, designs and composition. They have already been or are at present being used to treat a variety of different medical problems (4, 16, 20, 22–25, 27–29, 31–37, 43, 54, 56, 61, 96–102, 117, 118, 120–124, 126–130) including Parkinson's disease, multiple sclerosis, cerebral palsy, idiopathic tremors, epilepsy, spinal cord or other neurological injuries, paralysis such as paraplegia and quadriplegia (especially chronic, relatively intractable) pain, depression, anxiety, schizophrenia, strokes, "locked-in" syndrome, etc. The last explicitly listed medical entity (syndrome) characteristically presents clinically as a condition in which a person (patient), generally/usually as a consequence of stroke(s), has lost all abilities to communicate thoughts, feelings, desires and needs to the surrounding environment, usually including every other person therein.

Consequently, it might be readily understandable that it might not be unconscionable to treat or to clinically/therapeutically minimize in stimulative ways, such as those referred to above, extreme, substantial or otherwise untreatable significant neurological-disease-related or injury-caused dementia or markedly disabling, deteriorative impairment, such as can be caused by trauma, multiple infarcts and cognitive-function-undermining disease processes of diverse kinds, including, conceivably at least, some aspects or symptoms of Alzheimer's disease. The preferable use of macroscopically, surgically, entirely noninvasive brain/neural stimulative devices/equipment might usefully entail: (a) electromagnetic fields/waves, (b) acoustic/sonic, e.g., ultrasonic fields/waves and/or (c) pharmaceutical/medicinal substances, some or all of which might be delivered in precisely focused ways, as conceptualized to be within the broad-based category of BSNP. Such approaches, as reported and reflected on in the above list of references, might reasonably be expected to yield clinically and therapeutically worthwhile results in terms of patients' responses.

The increasingly diverse research application of surgically noninvasive treatments (such as in relation to depression and anxiety disorders, including obsessive-compulsive disorder (4), for example, transcranial magnetic (brain) stimulation (TMS), repetitive or rapid-rate TMS, rTMS, cranial or, virtually equivalently, transcranial electrostimulation (CES or TES), suggest various modes of access to research and development of effectively therapeutic methodologies. Electromagnetic, acoustic (e.g., ultrasonic), pharmacologic and other kinds of stimuli (as noted below) might with suitable values of stimulus parameters and appropriate target foci, induce or contribute to the induction of reward or pleasure responses that might conceivably be of higher intensity than those generally associated with sexual and eating-linked indulgence and gratification (20, 23). And such stimuli might be found to be especially noteworthy and clinically valuable with respect to research and development of effective, remedial BSNP, especially in terms of the pleasurization/enjoyitization and de-effortization of learning and work(ing).

As suggested by the following quotations (a–f), examples of technologies that might be useful (4, 19, 34, 129, 133, 134) in the research and development of learning and work-skills facilitative (LWF) and other modes of helpful/therapeutic brain stimulation/neural prosthetic functional modifiers/pacemakers (BSNP) might include (perhaps, repetitive) transcranial magnetic (brain) stimulation (TMS or rTMS), vagal nerve stimulation (VNS), cranial or (also known as) transcranial electrical stimulation (CES or TES) (131) and deep brain stimulation (DBS) (132).

(a) On page 50 (34), "Left prefrontal rTMS was associated with a mildly improved mood and greater mental alertness."

(b) On page 57 (4), under experimental conditions different from those entailed in quotation (a), "Left prefrontal TMS results in slight increases in subjective sadness, whereas right prefrontal repetitive TMS (rTMS) causes increased happiness" (133).

(c) On page 57 (4), under experimental conditions different from those involved in quotation (b), "Left prefrontal rTMS has been observed to have antidepressant effects" (133).

(d) On page 58 (4), "TMS at specific (neuroanatomic) regions, (stimulus/stimulation) intensities and frequencies might serve as a new treatment option for a host of conditions and might have applications for enhancing or modifying normal functions such as memory or (employment-related/work) skill acquisition" (133).

(e) On page 70 (134), "A feeling of relaxation" was noted in association with left vagal stimulation.

(f) On page 58 (4), the reproducibly observed and noted "anticonvulsant action of vagal nerve stimulation (VNS)" has been adequately significant such that "Now VNS is FDA approved for the treatment of epilepsy and about 3,000 people in the U.S. have these generators implanted" (133). And continuing observations are such as to facilitate the inference that the device "offers hope to epileptics" (129).

(g) Also on page 58 (4), "VNS parameters can affect learning and memory" in facilitative ways (133) and (19) "could have far-reaching effects, such as enhancing memory or treating obesity by curbing appetite … because the vagus nerve sends messages to the brain" that can signify that the "stomach is full."

Mind-particle/consciousness-experiencer (MP/MPEX/MPEC or MP-CONEX) research (coincidentally, having research implications for BSNP) might begin with the highest possible resolution or detail-revealing imaging modality/ies being used in such ways as to focus on a circumscribed point or area (maybe a moving target or center) of maximal activation that may—or may not—be observed to travel around, throughout the brain, elsewhere in the nervous system and, conceivably, even elsewhere in the body, outside of the nervous system. Functional magnetic resonance imaging (fMRI), magnetoencephalography (MEEG or MEG) (38), and computer-analyzed electroencephalography (CEEG) might possibly be seen as (being at least among) the current technology/ies of relative choice with which to initiate a search for a mind (particle).

A research project centering on a search for a potentially perceptible MPEC might instrumentally involve selective applications and adaptations of technologies, principles, properties, processes and other phenomena at least as diverse as the following list (and, undoubtedly, more so): cytoskeletally associated microtubules (135) as might comprise a tubular, tunneling rapid transit system for mind particles; superposition principles, including both constructive and destructive interference; particle accelerators, particle detectors, photomultipliers, scintillation counters, cloud or bubble chambers; possibly "noncontact," contactless or contact-free ultrasound (i.e., "a noncontact ultrasound device that can work as far as two inches from the skin, [because] the investigators added layers of material to the sound emitter, thereby matching its impedance to that of air") (68); infrasound, acoustic/sonic adaptations of laser's principles (55), holographic principles, tomographic principles, shock-wave-like acoustic phenomena, for example, low-energy extracorporeal shock wave therapy (ESWT) (136), pulsed, variable, static or continuous waveforms, discrete or discontinuous waves, components of supersonic compression waves, hypersonic waves, echoencephalography, sonar, Doppler ultrasonography, three-dimensional (3D) ultrasonography, echolocation, diverse electromagnetic phenomena including intersecting laser beams, diamagnetic and antimagnetic materials, infrared spectroscopy, thermal infrared cameras, thermography, radar, transceivers, transponders or other kinds of transmitter-receivers, transducers, microwaves, ultraviolet light, Cerenkov radiation (involving electromagnetic waves that are analogous to supersonic compression waves), transcranial magnetic stimulation (TMS), repetitive TMS (rTMS), magnetic resonance imaging (MRI), functional MRI (fMRI), superconductivity, superfluidity, superconducting quantum interference devices (SQUIDs), Josephson junctions, magnetic levitation, the Meissner-Ochsenfeld effect, quantum tunneling, antimatter, positron emission tomography (PET) scanning, single photon emission computed tomography (SPECT) imaging, electron beam computed tomography (EBCT), electron beam analysis, digitization, analog/digital interconversion, genetic engineering of "smarter" mice, possibly involving N-methyl-D-aspartate (NMDA) receptors, medicinal substances injected or permeated into an individual's circulatory system that might be activated by sharply focused physical modalities, such as precisely targeted sonic/acoustic and/or electromagnetic phenomena (for example, lasers), only whenever and wherever they reach, permeate and engage in the process of passing through one or more therapeutically optimal sites in any cases of disease/disorder-afflicted BBPs (137); "smart pills" perhaps entailing chemicals that can infiltrate neurons and stimulate an overproduction of "memory protein (a form of CREB)," that is, "Cyclic AMP Response Element Building Protein" (138); "time-reversal acoustics" (41), i.e., sequence-reversal acoustics, focusing mirror-assemblies, reflectors, fiber optics, optical computing technologies, neural networks, very large-scale integrated circuits (VLSI), microcircuits, minimicrocircuits or nanocircuits, minimicroprocessors/nanoprocessors, centrifugal or centripetal forces, gravity and antigravity, ballistics, antiballistics, wind, solar, geothermal (139); hydrogen-derived (140) and other kinds of power/energy, electromyography (EMG), anatomic/neuroanatomic regional blood flow analyses, electrocorticography, computer-spectral-analyzed electroencephalography (CEEG), evoked potentials (EPs), "artificial neurons," (combining both digital and analog processing) "that either excite or inhibit each other," depending on "feedback from other neurons" (141); electroconvulsive therapy (ECT), neuro- or other bio-feedback, virtual reality, vagal or vagus nerve stimulation (VNS) (4), nerve regeneration, synapses, neurotransmitters, partial cloning of specialized body parts, neural and possibly other kinds of stem cells, tissue-, partial- and whole organ regrowth, re-specialization/-differentiation, "neuron transistors" (39, 43), field effect transistors (FETs) (62–67), nanotransistors including single-molecule transistors (69) (anti-angiogenesis or) angiogenesis-related methods (142, 143), alternative medicine such as acupuncture and herbalism, noninvasive, non-implanted superficial or deep brain stimulation, macroscopically-surgically noninvasive devices or mechanisms, quantum electronic devices, carbon and other kinds of nanotubes, DNA-(e.g., gene chips, that is, DNA-coated microchips) or other kinds of microchips or micro-arrays, phased arrays or beam antennas, osmosis, percutaneous or transdermal permeation, iontophoresis (76, 144), electrophoresis, ionic (molecular, atomic or subatomic) propulsion systems, molecular diodes, molecular ratchet motors, "smart" drugs or pharmacologic agents with built-in, engineered mechanisms by which to affect and interact only with specifically identified cells, cell receptor types, reflective interfaces between superconductors and normal conductors, particle and waveform beams of diverse kinds, interferometry, scanning sensors of various types, "biophotonics," that is, biologically/medically applied "photonics," which refers to a combination of light-entailing and electronic technologies, nanotechnologically microminiaturized, microscopic-sized bubbles (145–146) that transport discrete packets, packaged doses, streams of medicinal substances or biochemically/biophysically active and effective agents/modalities directed to any particular site(s) or structurally/functionally significant anatomic destinations, a way of "nudging" and potentially transporting individual atoms to specific anatomic/neuroanatomic sites, involving adaptations of scanning tunneling microscopes (147); neuromodulators, pacemakers/pacesetters and other (neuro-) prostheses of numerous different kinds, robots, microminiaturized pharmacologic and/ or electronic nanorobots/"nanobots" (39, 43), other means by which to enable and bring about transmission of therapeutic agents or stimuli in macroscopically nondisruptive ways, hence, effectively, noninvasively from anywhere outside of a patient's (or other category of help-seeker, such as student or work-skills-job-seeker's) body to anywhere, that is, any specific location(s) inside the person's body, etc.

This list, in terms of its actual literal contents as well as what else may be inferable from it, might barely begin to anticipate some of many different approaches and numerous yet-to-be-envisioned and devised methods that could possibly provide valid mechanistic vehicles or pathways of constructive potential, whereby it might be feasible to conduct and implement appropriate research, development and application of both BSNP/neural prostheses and MPEX-related phenomena. There may emerge many beneficial methodologies, modalities and practical applications that might facilitate fulfillment of either or both of these categories of potential (BSNP and MPEC circulation, MPEC-virtual-circulation-internet), which may be markedly strengthened and diversified by genetic strategies which, in and of themselves, seem to have reached significant levels of attainment of knowledge and seem to promise impressive heights of prospective quality-of-life-improving wherewithal and well-being.

Mind-particle-consciousness-experiencer-circulation / virtual-circulation-internet (MPEX-CVCI) might validly be anticipated to be several or many years or perhaps several decades away from a time when their research and development might possibly come to fruition and their potential promise might be fulfilled. In contrast with these probably protracted prospects, brain-stimulation-neural-prosthetic-pacemaker/pacesetter functional-modulation-modifiers might possibly be brought to substantial fruition at any time between the present moment and several years (conceivably, depending on the pace of relevant, appropriate research and development, less than ten years) from now. The notion that brain stimulation, neuroprosthetically-mediated-learning and work-skills facilitation (BSNP-LWF) might already be technologically fully feasible seems significantly supported by the emerging BSNP-type or BSNP-like treatment modalities that are already being assessed and used (albeit in early phases of research and development) as therapeutic modalities for (symptoms of) Parkinson's disease, multiple sclerosis, idiopathic tremor, epilepsy, cerebral palsy, otherwise untreatable bodily pain, anxiety, depression, etc.

BSNP, genetic therapeutic methods, and MP-CONEX circulation/virtual-circulation-internet might help to ensure that, in the not-very-distant future, or (at any rate) eventually, virtually everyone might be broadly, deeply and diversely well educated. And the following quotation from a recent article (148) seems to render transparently appreciable that it might behoove each individual (current and potential) patient to become medically (and, by extrapolation, if not by implication, also otherwise) well educated.

"The most underused resource in the health care system is the patient's time. I have
begun suggesting that my patients look up their diseases on the Internet, (thereby)

learning about therapeutic options, drug side effects, necessary testing and monitoring. Compared to me, they have much more time and a greater incentive to research their disease(s), especially the particulars specific to them. I tell them they will soon know more than I do, and that that is good."

References

1. Andres J. C., Director, Family Practice Residency Training Program, Niagara Falls Memorial Medical Center, Niagara Falls, NY; personal communication: late 1999–early 2000.

2. Spangler R. A., Assoc. Prof., State Univ. of NY at Buffalo, Depts. of Physiology and Biophysics: personal communication, late 1999.

3. Butler S. R., Giaquinto S. Technical note: stimulation triggered automatically by electrophysiological events. *Med & Biol Engng*, now, *Med Biol Eng Comput*, 1969; 7: 329–331.

4. George M. S., Nahas Z., Lomarov M., Bohning D. E., Kellner C. How knowledge of regional brain dysfunction in depression will enable new somatic treatments in the next millennium. *CNS Spectrums* 1999; 4: 53–61.

5. Gevins A. S., Morgan N. H., Bressler S. L. et al., Human neuroelectric patterns predict performance accuracy. *Science* 1987; 235: 580–585.

6. Williamson S. J. How quickly we forget – magnetic fields reveal a hierarchy of memory lifetimes in the human brain. *Science Spectra* 1999; 15: 68–73.

7. Walgate J., Wagner A., Buckner R. L., Schacter D., Sharpe K., Floyd C. Memories are made of this. *Science & Spirit* 1999; 10, 1:7.

8. Connor S. Thanks for the memory. *The World in 1999*. The Economist Publications, 1999: 110–111.

9. Goetinck S. Different brain areas linked to memorization. *The Buffalo News*, final edn., Sun., June 13, 1999, Science Notes: H-6.

10. Hall S. S. Journey to the center of my mind, brain scans can locate the home of memory and the land of language. They may eventually help to map consciousness. *The New York Times Magazine*, June 6, 1999; section 6: 122–125.

11. Neergaard L. Studies shed new light on memory/studies take close look at brain's memory process. *Buffalo News* 1999; Fri., Aug. 21: A-10.

12. Sullivan M. M. (Editor). Task-juggling region in brain pinpointed. *Buffalo News* 1999; Sun., May 23: H-6.

13. Fox M. Test on rats turns thought into action. *Buffalo News* 1999; Sun., June 27: H-6.

14. McCrone J. States of mind, learning a task takes far more brainpower than repeating it once it's become a habit, could the difference show us where consciousness lies, asks J. M. *New Scientist*, March 20, 1999; 161: 30–33.

15. Pinker S. Will the mind figure out how the brain works? *Time* 2000; 155: 90–91.

16. Jacques S. Brain stimulation and reward: "pleasure centers" after twenty-five years. *Neurosurg* 1979; 5: 277–283.

17. Heath R. G. Modulation of emotion with a brain pacemaker. *Journal of Nervous and Mental Disease* 1977; 165: 300–316.

18. Langford K. H. of the Univ. at Birmingham, Dept. of Surgery/Neurosurgery, Birmingham, Ala., USA, personal communication, Aug. 10, 1987.

19. Neergaard L. Brain pacemaker helps treat depression. *Buffalo News* 1999; Tues., Oct. 12: A-4.

20. Heath R. G. Pleasure response of human subjects to direct stimulation of the brain: physiologic and psychodynamic considerations. *The Role of Pleasure in Behavior*: Harper & Row, 1964: 219–243.

21. Wanecski E. J. E., 1988 graduate of State Univ. at Buffalo/Niagara County Community College, co-sponsored electroencephalography technology training program and recipient of 1988 Graphic Controls Corp. award for EEG tech. excellence, personal communication, Mon., March 27, 2000. Also, please see ref. 22, which was received from/via E. J. E. Wanecski.

22. Callinan T. E. (edtr). Relief believed ahead for schizophrenia. *Rochester Democrat and Chronicle*, March 24, 2000; 168: 2A.

23. Bishop M. P., Elder S. T., Heath R. G. Attempted control of operant behavior in man with intracranial self-stimulation. *The Role of Pleasure in Behavior*: Harper & Row, 1964: 55–81.

24. Sem-Jacobsen C. W. Effects of electrical stimulation on the human brain. *Electroencephalogr Clin Neurophysiol* 1959; 11, 379.

25. Olds J. Pleasure centers in the brain. *Sci Am*, 1956; 193: 105–116.

26. Mancini L. Brain stimulation to treat mental illness and enhance human learning, creativity, performance, altruism and defenses against suffering. *Med Hypotheses* 1986; 21: 209–219.

27. Barker A. T., Freeston I. L., Jalinous R., Merton P. A., Morton H. B. Magnetic stimulation of the human brain. *J Physiol* 1985; 369: 3P.

28. Barker A. T., Freeston I. L., Jalinous R., Jarratt J. A. Magnetic stimulation of the human brain and peripheral nervous system: an introduction and the results of an initial clinical evaluation. *Neurosurg* 1987; 20: 100–109.

29. Bickford R. G., Guidi M., Fortesque P., Swenson M. Magnetic stimulation of human peripheral nerve and brain: response enhancement by combined magnetoelectrical technique. *Neurosurg* 1987; 20: 110–116.

30. Ebenbichler G. R., Erdogmus C. B., Resch K. L., et al. Ultrasound therapy for calcific tendinitis of the shoulder. *N Engl J Med* 1999; 340: 1533–1538.

31. George M. S. Brain activation involving mood and mood disorders, transcranial magnetic stimulation. *Audio-Digest Psychiatry* 1995; 24: Side B.

32. Talan J. Personal magnetism. Experimental method of treating depression helps 2 patients improve. *Newsday*, Tues., Nov. 21, 1995: B23, B26.

33. George M. S., Speer A. M., Wassermann E. M., et al. Repetitive TMS as a probe of mood in health and disease. *CNS Spectrums* 1997; 2: 39–44.

34. Greenberg B. D., McCann U. D., Benjamin J., Murphy D. L. Repetitive TMS as a probe in anxiety disorders: theoretical considerations and case reports. *CNS Spectrums* 1997; 2: 47–52.

35. Sherman C. Magnetic stimulation may offer ECT alternative. *Clinical Psychiatry News* 1998; April: 9–10.

36. Wassermann E. M. Repetitive transcranial magnetic stimulation: an introduction and overview. *CNS Spectrums* 1997; 2: 2 1–25.

37. Stein L., Belluzzi J. D., Ritter S., Wise C. D. Self-stimulation reward pathways: norepinephrine versus dopamine. *J Psychiatr Res* 1974; 11: 115–124.

38. Clayton J. Caught napping: depression, Parkinson's and obsessive-compulsive disorder may have a common cause. They could all be triggered when a tiny part of the brain dozes off. *New Scientist* 2000; 165: 42–45.

39. Kurzweil R. *The Age of Spiritual Machines: When Computers Exceed Human Intelligence.* Penguin Books, 1999: 52, 80, 120, 124, 127–128, 205, 220–221, 279, 300, 307–308, 313–314.

40. Damasio A. R. How the brain creates the mind. *Sci Am* 1999; 281: 112–117.

41. Fink M. Time-reversed acoustics. *Sci Am* 1999; 281: 91–97.

42. Rotman D. Will the real nanotech please stand up? *Technology Review, MIT's Magazine of Innovation* 1999; 102: 44–53.

43. Kurzweil R. Live forever. *Psychology Today* 2000; Feb.: 66–71.

44. Neergaard L. Researchers testing rub-on medicines: "skin enhancer" creams, gels may eventually replace ingested drugs. *Buffalo News* 1998; Tues., Sept. 29: A-6.

45. Charlier J. C. Tiny pipes with a big future. *Science Spectra* 1999; 17: 64–69.

46. Fry W. J. Electrical stimulation of brain localized without probes – theoretical analysis of a proposed method. *J. Acoust Soc Am* 1968; 44: 919–931.

47. Fry F. J. (brother of the late W. J. Fry; please note ref. no. 46), of the Indianapolis Center for Advanced Research, Indianapolis, Indiana, USA, personal communication, May 27, 1987.

48. Barker A. T. of the Department of Medical Physics & Clinical Engineering at Sheffield University, Sheffield, England, personal communication, Jan. 8, 1990.

49. Spangler R. A. of the Department of Biophysical Science and Physiology, School of Medicine and Biomedical Sciences, State University of New York at Buffalo, Buffalo, New York, USA, personal communication, Oct. 18, 1991.

50. Fink M., Prada C. Ultrasonic focusing with time-reversal mirrors. *Advances in Acoustic Microscopy Series.* Edited by A. Briggs and W. Arnold. Plenum Press, 1996.

51. Fink M. Time-reversed acoustics. *Physics Today* 1997; 50: 34–40.

52. Kuperman W. A., Hodgkiss W., Song H. C., Akal T., Ferla C., Jackson D. R. Phase conjugation in the ocean: experimental demonstration of an acoustic time-reversal mirror. *J Acoust Soc Am* 1997; 102: 1–16.

53. Fink M. Ultrasound puts materials to the test. *Physics World* 1998; 11: 41–45.

54. Rinaldi P. C., Jones J. P., Reines F., Price L. R. Modification by focused ultrasound pulses of electrically evoked responses from an in vitro hippocampal preparation. *Brain Res* 1991; 558: 36–42.

55. Watson A. Pump up the volume, what lasers do for light, sasers promise to do for sound – once you can work out the best way to build one. *New Scientist*, March 27, 1999; 161: 36–40.

56. Oldham J. Thoughts control a computer. *Popular Mechanics* 1999; 176: 28.

57. Sullivan M. M. (Editor). Atom-size circuits envisioned. *Buffalo News* 1998; Sun., Aug. 30: H-6.

58. Vogel M. Big minds gather to think small, really small. *Buffalo News* 1998; Sat., Oct. 24: C-5.

59. Fox M. Test on rats turns thought into action. *Buffalo News* 1999; Sun., 6-27: H-6.

60. Kurzweil R. The coming merging of mind and machine: the accelerating pace of technological progress means that our intelligent creations will soon eclipse us – and that their creations will eventually eclipse them. *Sci Am* 1999; 10: 56–60.

61. Chase V. D. Mind over muscles: when two emerging technologies meet, paralyzed people can move their limbs – just by thinking about it. *Technology Review: MIT's Magazine of Innovation* 2000; 103: 38–41, 44–45.

62. Schatzthauer R., Fromherz P. Neuron-silicon junction with voltage-gated ionic currents. *European Journal of Neuroscience* 1998; 10: 1956–1962.

63. LaBar K. S., LeDoux J. E. Partial disruption of fear conditioning in rats with unilateral amygdala damage: correspondence with unilateral temporal lobectomy in humans. *Behavioral Neuroscience*; 110: 991–997.

64. Watson A. Why can't a computer be more like a brain? *Science* 1997; 277: 1934–1936.

65. Taubes G. After 50 years, self-replicating silicon. *Science* 1997; 277: 1936.

66. Service R. F. Neurons and silicon get intimate. *Science* 1999; 284: 578–579.

67. Vassanelli S., Fromherz P. Transistor probes local potassium conductances in the adhesion region of cultured rat hippocampal neurons. *The Journal of Neuroscience* 1999; 19: 6767–6773.

68. Brown P. G. Q-bits, don't touch. *The Sciences* 2000; 40: 10.

69. Oldham J. (Ed.). One-molecule transistors. *Popular Mechanics* 2000; 177: 24.

70. Brooks M. Quantum clockwork. *New Scientist* 2000; 165: 28–31.

71. Zeilinger A. Quantum teleportation. *Sci Am* 2000; 282: 50–59.

72. Yerkes D. (Editor). *Webster's New Universal Unabridged Dictionary*. Barnes & Noble Books, Random House Publishing, Inc., 1996: 645–646, 1482.

73. Crystal D. (Editor). *The Cambridge Encyclopedia*, 2nd edn. Cambridge University Press, 1994: 1163.

74. Irvine M. Virtual reality changing the way doctors learn. *The Buffalo News*, June 18, 2000: H-6.

75. Davis H. L. Hand on the future: UB is developing virtual reality glove. *Buffalo News* 2000; Thurs., 6–29: B-1.

76. Walker P. M. (Editor). *Chambers Dictionary of Science and Technology*. Chambers Harrap Publishers, 1999: 877, 1190, 1240.

77. Dennett D. C. *Consciousness Explained*. Toronto, Little, Brown & Company, 1991; 101–102, 430.

78. Wade N. Brain may grow new cells daily. *The New York Times* 1999; Fri., Oct. 15:A-1 & A-21.

79. Talan J. The infinite mind. *Psychology Today* 1999; 32: 16.

80. Sinha G. Memory Expansion. *Popular Science* 2000; May: 44.

81. Hales D., Hales R. E. The brain's power to heal, major advances in the '90s: the most dazzling discovery is that the brain can generate new cells. *Buffalo News, Parade Magazine* 1999; Sun., Nov. 21: 10.

82. Squire L. R. *Memory and Brain*. Oxford University Press, 1987: 115–117.

83. Restak R. M. *The Brain*. Bantam Books, 1984: 245–269.

84. Gazzaniga M. S. The split brain revisited. *Sci Am* 1998; 279: 50–55.

85. Hamer D., Copeland P. *Living with Our Genes: why they matter more than you think*, 1st edn. New York: Doubleday, 1998: 18.

86. Davis H. WNY (Western New York) gene pool. *Buffalo News* 2000; Weds., 6-28: A-1, A-7.

87. Neergaard L. Apes' minds may shed light on evolution. *Buffalo News*, Weds., 6-28: A-1, A-7.

88. Wrangham R., Peterson D. *Demonic Males*. Boston: Houghton Mifflin Company, 1996: 40–41.

89. Wilson J. Miracles of the next 50 years. *Popular Mechanics* 2000; 177: 52–57.

90. Winston R. *Twins*. The Learning (television) Channel (TLC). British Broadcasting Corporation late May 1999.

91. Hall C. S., Nordby V. J. *A Primer of Jungian Psychology*. Mentor, Division of Penguin Books, 1973: 24–25, 38–41, 123.

92. Norton A. L., Ed. *Dictionary of Ideas*. London: Brockhampton Press, Helicon Publishing Ltd, 1994: 105.

93. Aspect A., Dalibard J., Roger G. Experimental test of Bell's inequalities using time-varying analyzers. *Physical Review Letters* 1982; 49: 1804–1807.

94. Coyle M. Time travel: a scientific possibility? *UFO Magazine*, 1999; 14: 50–55, 60.

95. Walker E. H. *The Physics of Consciousness: Quantum Minds and the Meaning of Life*. Cambridge, MA.: Perseus Books, 2000: 120–129.

96. Meyerson B. A., Boethius J., Caisson A. M. Percutaneous central gray stimulation for cancer pain. *Appl Neurophysiol* 1978; 41: 57–65.

97. Akil H., Richardson D. E., Hughes J., Barchas J. D. Enkephalin-like material elevated in ventricular cerebrospinal fluid of pain patients after analgesic focal stimulation. *Science* 1978; 201: 463–465.

98. Hosobuchi Y. Periaqueductal gray stimulation in humans produces analgesia accompanied by elevation of beta-endorphin and ACTH in ventricular CSF. *Modern Problems in Pharmacopsychiatry* 1981; 17: 109–122.

99. Dieckmann G., Witzmann A. Initial and long-term results of deep brain stimulation for chronic intractable pain. *Appl Neurophysiol* 1982; 45: 167–172.

100. Tsubokawa T., Yamamoto T., Katayama Y., Hirayama T., Sibuya H. Thalamic relay nucleus stimulation for relief of intractable pain. Clinical results and B-endorphin immunoreactivity in the cerebrospinal fluid. *Pain* 1984; 18: 115–126.

101. Young R. F., Kroening R., Fulton W., Feldman R. A., Chambi I. Electrical stimulation of the brain in treatment of chronic pain. Experience over 5 years. *J Neurosurg* 1985; 62: 389–396.

102. Blumenkopf B. Chronic pain relief with deep brain stimulation. *The Psychiatric Times/Medicine & Behavior* 1988; Sept. 8–9.

103. Thomas K. R., Folger K. R.; Capecchi M. R. High frequency targeting of genes to specific sites in the mammalian genome. *Cell* 1986; 44: 419–428.

104. Mancini L. Riley-Day syndrome, brain stimulation and the genetic engineering of a world without pain. *Med Hypotheses* 1990; 31: 201–207.

105. Stover, D. Growing hearts from scratch. *Popular Science* 2000; 256: 46–50.

106. Mooney D. J., Milos A. G. Growing new organs. *Sci Am* 1999; 280: 60–65.

107. Recer P. Neural stem cells in mice are found to grow organ, muscle, other tissues. *Buffalo News* 2000; Fri., 6-2: A-6.

108. Howland D. Organs grown using cells from animals. *Buffalo News* 1997; Weds., 7-23: A-1, A-4.

109. Petranek S. L. (Editor). Growing organs. *Discover* 2000; 21: 112–113.

110. Bilger B. Metamorphoses: hair-raising feats in cell biology. *The Sciences* 1999; 39: 6–7.

111. Friedman R., Ed. *The Life Millennium, The 100 Most Important Events & People of the Past 1,000 Years*: Life Books, Time Inc., 1998: 188.

112. Keesling B. Beyond orgasmatron. *Psychology Today* 1999; 32: 58–60, 62, 84–85.

113. Carpenter M. B. *Core Text of Neuroanatomy*. Baltimore: Williams & Wilkins Co., 1972: 74, 94, 180.

114. Zorpette G. Muscular again: within a decade or two, scientists will create a genetic vaccine that increases muscle mass – without exercise. *Sci Am* 1999; 10: 27–31.

115. Hoffmann-Wadhwa N. I., 1981 graduate of St. George's Univ. Schl of Medicine, Grenada, West Indies, personal communication, Thurs., June 15, 2000.

116. Stensaas S., Stensaas L., Depts. of Anatomy and Physiology, Univ. of Utah, Salt Lake City, Utah, personal communication, Mon., June 19, 2000.

117. Ritter M. Blind man navigates whole new world of vision by using tiny camera wired directly to his brain. *Buffalo News* 2000; Mon., Jan 17: A-14.

118. Mead C. *Analog VLSI and Neural Systems*. Addison-Wesley Publishing Co. Inc. 1989: 229, 257, 279, 294.

119. Peterson S. Ultrasound breakthrough offers hope blind may see. *Buffalo News* 1992; Weds., Jan. 15: A-1.

120. Kwiatkowski J. Sweet sounds: breaking into the new world of hearing with the sometimes controversial cochlear implant. *Buffalo News* 2000; Tues., Feb. 22: C-1, C-2.

121. Fischer J. S. Tweaking nature's repair kit: one's own cells may be the best medicine. *U.S. News & World Report* 2000; 128: 55.

122. Brenner M. J. The new Atlantis and the frontiers of Medicine. *JAMA* 2000; 283: 2296.

123. Editorial staff. Quadriplegic students may be walking before this summer's sunshine ends. *Medical World News* 1982; 23: 58–59.

124. Brand C. Computer chip helps paralyzed man to walk: paraplegic walks, thanks to computer implant. *Buffalo News* 2000; Tues., 3-21: A-1, A-2.

125. Sullivan M. M. (Editor). Tidy line of molecules could mean much to science. *Buffalo News* 1997; Sun., Feb. 16: H-6.

126. Friend T. Scientists pinpoint brain's "sleep switch." *USA Today* 1996; Fri./Sat./Sun., Jan. 12–14: 1-A, l-D.

127. Moon M. A. Electrical stimulation tempers advanced Parkinson's, subthalamic nucleus implants. *Internal Medicine News, Clinical Rounds* 1999; Jan.: 14.

128. Douma A. Cause of tremors remains unknown. *Buffalo News* 1999; Fri., 10-29: B-15.

129. Davis H. L. Implant offers hope to epileptics. *Buffalo News* 1999; Mon., 3-8: B-1.

130. Sullivan M. M. (Editor). System helps paralyzed write using brain waves. *Buffalo News* 1999: Thurs., 3-25: A-4.

131. Stinus L., Auriacombe M., Tignol J., Limoge A., Le Moal M. Transcranial electrical stimulation with high frequency intermittent current (Limoge's) potentiates opiate-induced analgesia: blind studies. *Pain* 1990; 42: 351–363.

132. Andy O., Jurko F. Thalamic stimulation effects on reactive depression. *App Neurophysiol* 1987; 50: 324–329.

133. Privitera M. R., Clinical Associate Professor, Dept. of Psychiatry, Strong Memorial Hospital, Mood Disorders Center, Rochester, New York 14627, personal communication, 8/16/99.

134. Upton A. R. M., Tougas G., Talalla A. et al. Neurophysiological effects of left vagal stimulation in man. *Pace* 1991; 14: 70.

135. Penrose R., Shimony A., Cartwright N., Hawking S. *The Large, the Small and the Human Mind*. Cambridge University Press 1997, paperback 1999: 128–143.

136. Douma A. Painful bone spurs gradually go away. *Buffalo News*, Sat., June 17, 2000: B-21.

137. Tsien J. Z. Building a brainier mouse. *Sci Am* 2000; 282: 62–68.

138. Weed W. S. Smart pills: how about a little Viagra for your memory? *Discover* 2000; 21: 82.

139. Sullivan M. M. (Editor). A crisis of will, not energy. The Buffalo News 2000; April 3: B-2.

140. Sullivan M. M. (Editor). New way to make hydrogen gas seen promising. *The Buffalo News* 2000; Sept.: H-6.

141. Fordahl M. It's not Hal but new circuit mimics the way the brain works. *Buffalo News*, Thurs., June 22, 2000: A-l, A-4.

142. Folkman M. J. Can Conventional Anti-Cancer Therapies Be Improved by Inhibition of Angiogenesis? Centennial Symposium Cancer Genetics & Biology, Oct. 9, 1998. Roswell Park Cancer Institute, Buffalo, New York, USA. Re: M. J. Folkman, Andrus Professor of Pediatric Surgery & Professor of Cell Biology, Harvard Medical School, Children's Hospital, Boston, MA.

143. Hanrahan D. Induction of Angiogenesis and Acquired Resistance to Apoptosis along Pathways of Multistep Tumorigenesis. Centennial Symposium Cancer Genetics & Biology, Oct. 9, 1998. Roswell Park Cancer Institute, Buffalo, New York, USA. Re: D. Hanrahan: Dept. of Biochemistry & Biophysics, Hormone Research Institute, Univ. of California, San Francisco, CA.

144. Douma A. Stress can aggravate excessive sweating. *The Buffalo News*, 2000. Sat., June 10: B-1.

145. Davis H. L. Shedding light on the future. *Buffalo News*, Sun., April 2, 2000: A-1, A-12.

146. Prasad P. N., of the Depts. of Chemistry and Physics and the Institute for Lasers, Photonics (optical technologies) and Biophotonics at the State U. of N.Y. at Buffalo, N.Y., USA, personal communication, Oct. 25, 1999.

147. Sullivan M. M. (Ed.). A new way to nudge individual atoms. *The Buffalo News*, Sun., April 23, 2000: H-6.

148. Merrill M. Overworked hospital staff should learn to put patients to work. *The Buffalo News*, June 25, 2000: H-1.

conclusion

Within the next 1 to 100 plus years, learning and "working" will become virtually effortless, instantaneous and automatic via mind and brain stimulation that will be mediated as follows. Pleasurable mind-particle (MP) and brain stimulation will be delivered to the learning and "working" brain and body whenever and only whenever the MP, brain and body are emitting physiological phenomena (for example, brainwaves in the 12 to 42+ Hz or cycles per second range, refs. 1 and 2) that indicate that the mind, brain and body are engaged in learning and/or working (i.e., focused activity, perception, learning and memory formation). Work will become so much fun and so quick to accomplish that it will no longer seem like effortful "work."

Reproducible out-of-body experiences, global positioning systems and global MP circulation, mediated by stimulation of various sites in the human brain (refs. 3–5) will enable every human earthling to be equally well acquainted with every other earthling in an altogether friendly context. Hence, every earthly human being who has ever lived (or will ever live) will be equally famous among all the other humans who will have ever lived. And each human who has ever been (or will ever be) born on Earth is already and will always be infinitely famous among the universe's infinite population of Heaven-dwelling mind/souls.

Eventually, perhaps within the next 50 to 100+ years, once all "yet-to-live" earthlings have lived and "died," then all human and other earthling bodies (including brains) will be abandoned altogether as the MPs of all currently "alive," i.e., present (and all future) earthly organisms harmoniously join ranks and merge with the aggregate population of all MPs of "deceased" earthlings. Then all past, present and future earthling-MPs/MSPs will collectively comprise one big, happy Big Bang Family of mind-soul-particles as was intended by the President of the Universe (God?) at the time of the last, most recent Big Bang Explosion and concomitant selection of planet Earth as the current cosmic central-reference-point.

Life on Earth is basically "Hell." Life surrounding planet Earth is basically "Heaven," except for a finite zone of Purgatory that is spatially intermediate between Earth and spatially infinite true Heaven. Within the next 50 to 100+ years, Hell (Earth) and Purgatory (where the deceased and yet-to-be born earthlings now dwell together) will be de-Hellified and de-Purgatized, hence turned into true Heaven and the three modes of existence will merge into a single, Heavenly one.

Once Hell and Purgatory have been fully Heavenized, there will be a new Big Bang Explosion when the President of the Universe (God?) selects a new central-reference-point (CRP) planet which will also be the new "Hell." So, the cycle of de-Hellification, de-Purgatization and Heavenization (i.e., a Big Bang followed by gradual [1] scientific, technological and philosophical improvement of initially horrible conditions on the newly-selected Hell-CRP planet; some planet other than burned-out, expendable Earth) will begin, proceed and progress all over again.

It is worth pointing out that despite our human tendency to think that scientific and technological progress have no upper limits, Heaven is virtually definable as the infinitely large volume of space, that is, an infinitely large place (with Hell and its peripheral, finite Purgatory at Heaven's center) wherein science and technology are maximal and optimal, so that every wishful thought and every wildest and fondest "dream" are fulfillable there and no kind of suffering is possible there. Moreover, since the universe operates on the pleasure principle, it stands to reason that, outside of CRP-Hell (planet Earth), wishful thinking is accurate thinking. And without CRP-Hell, the universe could not exist at all; at least, not as a basically orderly place.

References

1. Fox D. Brainwave boogie-woogie. *New Scientist*, Dec. 24/31, 2005; 188 (2531/2532): 50–51.
2. Biello D. Searching for God in the brain. *Scientific American Mind*, Oct./Nov. 2007; 18 (5): 38–45.
3. Kotler S. Extreme states. *Discover*, July 2005; 26 (7): 60–67.
4. Hoppe C. Controlling epilepsy. *Scientific American Mind*, June/July 2006; 17 (3): 62–67.
5. Bosveld J. Soul search: can science ever decipher the secrets of the human soul? *Discover*, June 2007; special issue: 46–50.

About the Author

The author, born in 1951, is a physician who was in training to become a psychiatrist when he became disabled with severe obsessive-compulsive disorder (OCD) in 1985, which was the last year in which he was gainfully employed. He was at the end of the second year of a four-year psychiatrist's training program at that time. And, therefore, he is in possession of a two-year certificate of residency training. He has also been diagnosed with bipolar manic-depression since 2007 and had a manic "nervous breakdown" for which he was hospitalized for three weeks in February 2008. He became divorced back in 1985, and has a daughter who was born in 1983.

He has a two-year technician's degree in electroencephalography (that is, EEG or "brain-wave") technology and additional background in bioengineering and biophysics. He has a strong interest in astrophysics/cosmology, religion and the interface between science and religion.

Although he has been a fairly strong student with a background of 17 years of college, graduate school and personal education, he has been afflicted since childhood with an attention deficit disorder (ADD) and two different learning disabilities:

(1) gadgetaphobia, including computer illiteracy and
(2) general informational phobia.

And he has suffered from OCD since the age of 12. The computer-illiteracy problem might end when and if he can find someone to *patiently* teach him, in a one-to-one, "hands-on" way, how to use a computer.

Despite his problems, he is a graduate of St. George's University School of Medicine (1983) in Grenada, the West Indies. Moreover, in 1974, he worked briefly, as a research assistant, on "Artificial Vision for the Blind," under the auspices of Dr. William H. Dobelle in the Neuroprostheses and Artificial Organs Division of the Bioengineering Department at the University of Utah, in Salt Lake City, Utah. Dr. Dobelle is cited in the 2005 *Guinness Book of World Records* under the headings of "Medical Phenomena" and the "earliest successful artificial eye," on page 20.

The author won an award for "Academic and Technical Excellence" in EEG technology from Graphic Controls Corporation in 1988. He has had two articles published in *Speculations in Science and Technology* and four articles published in *Medical Hypotheses*. Three of these articles deal primarily with the question of how learning abilities and work skills could possibly be improved by (preferably, surgically noninvasive) brain stimulation.

One of his goals is to play whatever role he can in the conceivable implementation of brain-stimulation-mediated learning facilitation (LF) and work skills facilitation (WF), enhancement and diversification. And he has fairly specific ideas about how these goals could be accomplished.

References

1. Mancini L. S. How learning ability might be improved by brain stimulation. *Speculations in Science and Technology*, 1982; 5 (1): 51–53. (Reprint of this article included herein.)

2. Mancini L. S. Brain stimulation to treat mental illness and enhance human learning, creativity, performance, altruism, and defenses against suffering. *Medical Hypotheses*, 1986; 21: 209–219. (Reprint of this article included herein.)

3. Mancini L. S. Riley-Day Syndrome, brain stimulation and the genetic engineering of a world without pain. *Medical Hypotheses*, 1990; 31: 201–207. (Reprint of this article included herein.)

4. Mancini L. S. Ultrasonic antidepressant therapy might be more effective than electroconvulsive therapy (ECT) in treating severe depression. *Medical Hypotheses*, 1992; 38: 350–351. (Reprint of this article included herein.)

5. Mancini L. S. A magnetic choke-saver might relieve choking. *Medical Hypotheses*, 1992; 38: 349. (Reprint of this article included herein.)

6. Mancini L. S. A proposed method of pleasure-inducing biofeedback using ultrasound stimulation of brain structures to enhance selected EEG states. *Speculations in Science and Technology*, 1993; 16 (1): 78–79. (Reprint of this article included herein.)

7. Mancini L. S. (written under the pseudonym Nemo T. Noone). Waiting hopefully. *Western New York Mental Health World*, 1995; 3 (4), Winter: 14. (Reprint of this article included herein.)

APPENDIX

Medical Hypotheses 21: 209–219, 1986

BRAIN STIMULATION TO TREAT MENTAL ILLNESS AND ENHANCE HUMAN LEARNING, CREATIVITY, PERFORMANCE, ALTRUISM, AND DEFENSES AGAINST SUFFERING

Lewis Mancini

ABSTRACT

Any mental/emotional state or process (MESP) which is considered (a) highly desirable (e.g., sustained concentration, memorization of important facts, empathy) or (b) undesirable (e.g., paranoid delusionalism, delirium) could be, respectively, (a) facilitated or (b) deterred by means of an external (i.e., extracranial, or at least extracerebral, and extracorporal) brain stimulation circuit designed in such a way as to deliver rewarding stimulation as often and only as often as and for as long and only for as long as an electroencephalographic or other kind of brain function characteristic, which uniquely identifies the occurrence of the MESP in question, were being emitted by the individual's (i.e., the subject's) brain, with the intensity of the stimulation at every point in time being proportional, respectively, (a) to the simultaneous magnitude or (b) to the <u>reciprocal</u> of the simultaneous magnitude of the MESP-identifying characteristic. Approaches a and b are generalized examples of a number of hypothetical stimulation paradigms presented below that might be used to treat mental illness, enhance learning, etc. (as in the title). Explanations of the psychodynamic mechanisms whereby these paradigms might exert their intended effects are given in most cases.

Introduction

The methodology herein proposed is predicated on the inference that can be drawn from substantial experimental evidence (1, 2, 3, 4, 5, 6, 7) that any given mental/emotional state or process that one might want to either induce or suppress has characteristically and uniquely associated, detectable electroencephalographic and other kinds of pleasurable brain-function features (and a corresponding underlying, uniquely characteristic configuration of excited and inhibited neuroanatomic circuits) which could be used by an external circuit to automatically detect the occurrence of that MESP. Such features, which are characteristically linked with a particular MESP, may be thought of as linked characteristics (LCs) of that MESP. This methodology would require merely that each person's LCs be essentially time-invariant for her or him so that, for example, whenever person A engages in high-level concentration (MESP-CONC), A's brain reliably emits one or more particular LCs (LC-CONCs). It would make no difference whatsoever whether A's LCs were completely different from anyone else's or not.

What is meant by stimulation is the production or suppression of impulses or action potentials within minute volumes of brain tissue, for example, a sphere or "focal spot" with a diameter of l mm (8) or less, without any direct production or suppression of action potentials in surrounding tissue. For any given application in any given case, the stimulation might consist in a continuous waveform or (more probably) in successive discrete waveforms, that is, pulses. The technical implementation of such a system might entail the use of electroencephalographic, ultrasonic, and/or electromagnetic techniques, such as MRI, for LC determination, detection, and magnitude monitoring. According to Brown and Kneeland (9, p. 495), "More powerful magnets offer the possibility of monitoring phosphorus[31] and therefore cell energy metabolism" and therefore the possibility of monitoring MESP-specific cerebral activity as reflected in LCs. Stimulation that would be nondestructive might be affected either invasively (but hopefully not), e.g., via surgically implanted electrodes, or (preferably) noninvasively or minimally invasively (e.g., invasively with respect to the skull, such as with an ultrasonic irradiator implanted therein, but noninvasively with respect to the brain) by means of focused electromagnetism and/or ultrasound as is discussed in some detail in another paper (10).

The types of brain stimulation that might be used are (a) pleasurable (PLE), i.e., rewarding, (b) sedating (SED), (c) alerting (ALERT), (d) specific-MESP-excitatory (SMESPEX), (e) specific-MESP-inhibitory (SMESPIN), and (f) otherwise characterizable. Stimulation could be any one of, more than one of, or even all of the above in nature, with its nature in every instance being determined both by the particular neuroanatomic sites(s) focused upon and the particular stimulation parameters used. Distressing, i.e., aversive stimulation should never be used.

The essential principle of the methodology, stated in the most general terms possible, is that, by means of an ideally wholly external prosthetic system, the individual would receive brain stimulation of one or more kinds as often as and only as often as and for as long as and only for as long as (and with intensity either directly or inversely proportional to the magnitude with which) the individual would emit a predetermined LC and, hence, would be either facilitated or inhibited with respect to indulgence in the MESP corresponding to that LC. Hence, the delivery and the intensity of the brain stimulation would be <u>dependent</u> upon the magnitude of one or more LCs and, hence, could be referred to as <u>linked characteristic-dependent brain stimulation (LCDBS)</u>. Each LCDBS system would consist essentially in:

1. An LC detection and magnitude-monitoring component.
2. An LC-magnitude/stimulation intensity proportionalizing circuit.
3. A stimulating component.

RESULTS OF SOME (NON-LCDBS) BRAIN STIMULATION EXPERIMENTS

Sem-Jacobsen (11, p. 379) reported that "We have been able to obtain feelings of comfort, relaxation, joy, and intense satisfaction ... In the ventromedial part of the frontal lobe, regions of pleasure and relaxation are lower and more internal than those mediating anxiety and irritation. The responses of relaxation and comfort obtained from stimulation of the frontal lobe are so intense that psychotic episodes have been broken up in less than one minute on several occasions ... Stimulation of the ventromedial part of the frontal lobe has a calming effect, as does stimulation of the central region of the temporal lobe."

Heath (12, p. 224) found that "With septal stimulation the patients brightened, looked more alert, and seemed to be more attentive to their environment during, and for at least a few minutes after, the period of stimulation ... Expressions of anguish, self-condemnation, and despair changed precipitously to expressions of optimism and elaborations of pleasant experiences, past and anticipated. Patients sometimes appeared better oriented; they could calculate more rapidly and, generally, more accurately than before stimulation. Memory and recall were enhanced or unchanged." Moan and Heath (13) also reported sexual feelings to be associated with self-stimulation of the septal region.

The observation of Higgins, Mahl, Delgado, and Hamlin (14, p. 418) that "after one stimulation ..." of the "frontotemporal" area, which they define in neuroanatomic detail, a young male subject "... said, without apparent anxiety, 'I'd like to be a girl,' " whereas "In the last interview, when he came close to expressing a similar idea under pressure by the interviewer but in the absence of stimulation, he became markedly anxious and defensive" suggest that stimulation of that area could be useful in the treatment of the personality disorders intrinsic to which is the ostensible incorrigibility of maladaptive defense mechanisms. Hence, the neuroanatomic explored by these investigators might include appropriate stimulation sites for the applications discussed herein. The septal region in particular may contain utilizable sites (15, 16).

LCDBS AS TREATMENT FOR MENTAL ILLNESS

Examples of paradigms or modi operandi of LCDBS systems designed to treat mental illness and/or affect prophylaxis for antisocial, including criminal behavior, are as follows.

1. Treatment of Mental Illness Type I, that is, <u>psychosis</u> (including schizophrenia), which may be defined here simply as a phenomenon consisting in gross disorganization of mental processes and/or distorted perception of reality:

PLE stim. int. inc. in prop. as mag. LC-PSYCH dec.
and
PLE stim. int. dec. in prop. as mag. LC-PSYCH inc.

where PLE stim. int. = (i.e., abbreviates) pleasurable stimulation intensity, inc. = increases, dec. = decreases, in prop. as = in proportion as, mag. = magnitude of, and LC-PSYCH = an LC which identifies the occurrence of a psychotic process or state entailing, for example, hallucinations and/or delusions, and/or looseness of associations. Hence, the less psychotic an individual's mentation would become, the more intense the pleasure which he or she would obtain from the system (and the more psychotic, the less pleasure), so that the individual would be strongly motivated to discard psychotic processes and states. Antipsychotic SMESPIN stimulation paradigms might entail the use of inhibitory stimulation directed at some of the same sites in the brain as where antipsychotic drugs exert their therapeutic effects (17).

2. Treatment of Mental Illness Type II, that is, <u>mental pain</u> (MP), including all forms of neurosis. The three basic kinds of mental pain: (a) anxiety (grading up to terror), (b) depression grading up to hopelessness), and (c) anger (grading up to rage) pervade all forms of mental illness, especially the anxiety, depressive, personality, and adjustment disorders.

PLE stim. int. inc. in prop. as mag. LC-MP inc.
and
PLE stim. int. dec. in prop. as mag. LC-MP dec.

where LC-MP = an LC which identifies the occurrence of one or more kinds of mental pain. The rationale of this modus operandi would be that, inasmuch as the brain's reward (i.e., pleasure) and punishment (i.e., aversion or pain) systems are reciprocally inhibitory with respect to each other (18), the more intense the mental pain a person is experiencing, the more intense would be the pleasure needed to nullify, that is, inhibit, suppress, or relieve that pain. And the less intense the pain, the less intense the pleasure needed to nullify it. Hence, by dint of this modus operandi, mental pain would be automatically nullified by an intensity of pleasurable stimulation commensurate with it.

3. Treatment of Mental Illness Type III, that is, <u>maladaptive</u> and/or <u>destructive pleasure</u> (MDP), such as constitutes the motivational or affective essence of drug (including alcohol) abuse, sadism, some forms of sexual deviance, and mania:

SMESPIN-MDP stim. int. inc. in prop. as mag. LC-MDP inc.
and
SMESPIN-MDP stim. int. dec. in prop. as mag. LC-MDP dec.

where SMESPIN-MDP stim. = stimulation of one or more neuroanatomic sites which specifically causes inhibition of a particular kind of MDP and LC-MDP = an LC which specifically characterizes that kind of MDP. Hence, the greater the intensity of MDP (and, correspondingly, the greater the magnitude of LC-MDP), the higher the intensity of MDP-inhibitory (i.e., SMESPIN-MDP) stimulation that would be needed and automatically delivered to nullify it.

4. Treatment of Mental Illness Type IV, that is, pathological unconcern for others; facilitation of <u>empathy</u>, <u>compassion</u>, and <u>altruism</u> (ECA):

SMESPEX-ECA stim. int. inc. in prop. as mag. LC-ECA dec.
and
SMESPEX-ECA stim. int. dec. in prop. as mag. LC-ECA inc.

where SMESPEX-ECA stim. = stimulation of one or more neuroanatomic sites which specifically causes excitation of ECA toward others and LC-ECA = an LC which specifically characterizes the process or state of being empathic and/or compassionate and/or altruistic. Hence, the lower the intensity of an individual's ECA (and, correspondingly, the lower the magnitude of LC-ECA) and, correspondingly, the greater the degree of appropriateness of an increase in the intensity of that individual's ECA, the higher the intensity of ECA-excitatory (i.e., SMESPEX-ECA) stimulation that would be automatically delivered in order to affect that increase. And the greater the intensity of spontaneous ECA, and the less the degree of appropriateness of an increase in ECA, the less SMESPEX-ECA stimulation would be delivered. One or more of approaches 1-through-4-like approaches would be appropriate for prophylaxis of antisocial, including criminal, behavior.

In cases in which the rudiments of and hence the potential for ECA were so pervasively lacking as to be noninducible by virtually any means, then it might be possible to affect prophylaxis of antisocial behavior by means of what could amount to <u>arresting</u> LCDBS affected by implementation of the phenomenon of cortical suppression (19), that is, suppression of spontaneous electrical activity of cortical area 4 (i.e., the motor cortex) and hence suppression of all movement by dint of the stimulation of specific suppressor areas of the cerebral cortex, all of which have been demonstrated to project to the caudate nucleus. And inasmuch as when certain stimulation parameters were used, stimulation of the caudate in humans was found to be rewarding (20, 21), one might surmise that such movement-suppressive stimulation could be made both pleasurable and arresting. Whenever more than one LCDBS system were required, e.g., whenever criminal behavior is motivated by both sadism and anger or whenever a schizoaffective patient suffers from both active psychosis and depression, the two or more appropriate circuits would simply be operated in parallel with respect to each other.

EXAMPLES OF LCDBS SYSTEMS TO ENHANCE COGNITION AND PERFORMANCE

5. To improve learning ability or affect <u>learning facilitation</u> (LF) (22):

PLE stim. int. inc. in prop. as mag. LC-LTM inc.
and
PLE stim. int. dec. in prop. as mag. LC-LTM dec.

where LC-LTM = a linked characteristic which always and only occurs during the formation of a <u>long-term</u> <u>memory</u> <u>trace</u> (LTM) in the individual in question, i.e., the subject. If it were not readily possible to identify the occurrence of an LC-LTM, then a more generalized kind of <u>learning-linked</u> <u>characteristic</u> (LLC) such as an LC-CONC which always and only occurs in the subject in question during high-level, sustained <u>concentration</u>, which might more readily be identified, could be used in place of LC-LTM in this paradigm, by dint of which the process of long-term memory trace formation (or the state of high-level, sustained concentration or some other learning-linked MESP) would become intensely pleasurable for the subject and therefore likely to occur readily, rapidly, and sustainably. IF LCDBS methods would be based on the idea that if learning could be made intensely pleasurable, for example, at least as pleasurable as eating and sexual activity are for most of us after long periods of deprivation of these modes of gratification, we would be able to tip into what for most of us are our greatly underdeveloped intellectual potentials. Even our IQ scores might gradually rise, though probably not as rapidly or markedly as our actual demonstrable learning ability. In view of the realization that some individuals are far more nearly hedonistically optimized with respect to the learning process, that is, have far larger appetites for informational details than others do, it stands to reason that some individuals far more nearly attain to what may be considered their upper biological limits of intellectual functioning than others do. Hence, one would not necessarily predict a very high correlation between pre- and post-LF IQ scores. LF might prove to be of value not only for intellectually normal individuals but also as a treatment modality for the mentally retarded and for neurologically impaired individuals such as aphasic stroke victims (in particular vis-à-vis relearning language skills) and those afflicted with dementing processes such as Alzheimer's disease.

6. To create new interests or affect interest diversification in a person:

PLE stim. int. inc. in prop. as mag. LC-ATUN inc.
and
PLE stim. int. dec. in prop. as mag. LC-ATUN dec.

where LC-ATUN = a linked characteristic which characterizes the process of <u>attending</u> (AT--) to details of a kind, for example, details of car mechanics, robotics, or a foreign language, which would naturally, that is, without LCDBS be <u>uninteresting</u> (--UN) to the particular subject in question. Hence, by dint of this modus operandi, attending to and processing information of a kind which would otherwise bore the individual, would become intensely pleasurable and therefore likely to occur.

7. To enhance performance of skills or affect <u>performance enhancement</u> (PE):

PLE stim. int. inc. in prop. as mag. LC-METT inc.
and
PLE stim. int. dec. in prop. as mag. LC-METT dec.

where LC-METT = an LC which characterizes <u>meticulousness</u> (METT). Hence, the process of being meticulous would become intensely pleasurable and therefore likely to occur. By virtue of PE, working (like learning, by virtue of LF) could be rendered as pleasurable as eating and sexual activity are for most people. Consequently, productivity in the context of work might dramatically increase.

8. To enhance <u>creativity</u> (CR):

PLE stim. int. inc. in prop. as mag. LC-CR inc.
and
PLE stim. int. dec. in prop. as mag. LC-CR dec.
operated together with
SMESPIN-CRIN stim. int. inc. in prop. as mag. LC-CRIN inc.
and
SMESPIN-CRIN stim. int. dec. in prop. as mag. LC-CRIN dec.

where LC-CR = an LC which characterizes one or more creative processes such as might occur during dreaming when, unfortunately, under natural circumstances, that is, without brain stimulation, the individual's capability of converting concept into actuality is minimal, because he or she is immobilized by the sleep process. And SMESPIN-CRIN stim. = stimulation of neuroanatomic sites which specifically causes inhibition of these <u>creative</u> impulses which naturally occurs during the waking state in most people and is inherent in the perceptual and cognitive rigidity imposed by the conscious mind. And LC-CRIN = an LC which characterizes the naturally inhibited state of these impulses during the waking state.

<u>THE ABOLITION OF SUFFERING WITHOUT COMPROMISE OF ADAPTIVENESS</u>

The following example, entailing the simultaneous use of paradigms 9 through 11, suggests ways in which any and all suffering (e.g., anxiety, dyspnea, nausea) could be abolished without any compromise of its naturally associated adaptive value.

9. SMESPIN-POA/MP stim. int. inc. in prop. as mag. LC-POA/MP inc.
and

SMESPIN-POA/MP stim. int. dec. in prop. as mag. LC-POA/MP dec.
together with, i.e., operated in parallel with

10. PLE stim. int. dec. in prop. as mag. LC-POA/MP inc.
and
PLE stim. int. inc. in prop. as mag. LC-POA/MP dec.

11. SMESPEX-AVOID stim. int. inc. in prop. as mag. LC-POA/MP inc.
and
SMESPEX-AVOID stim. int. dec. in prop. as mag. LC-POA/MP dec.

where POA = <u>pain</u> and <u>other</u> forms of bodily <u>aversiveness</u>, SMESPIN-POA/MP stim. = stimulation of neuroanatomic sites which causes <u>inhibition</u> of <u>POA</u> and <u>MP</u>. And SMESPEX-AVOID stim. = stimulation of neuroanatomic sites which causes <u>excitation</u> of <u>avoidance</u> or withdrawal behavior. LC-POA/MP = an LC which characterizes one or more kinds of POA and/or MP. With LCDBS systems 9 through 11 operating in parallel with respect to each other, the more closely a person's body were to approach or be approached by a noxious (i.e., LC-POA/MP-inducing) stimulus, the more strongly his or her POA and/or MP would be inhibited by paradigm 9, so that he or she would experience no pain (23, 24, 25), other bodily aversiveness or mental pain, and the less pleasure he or she would be gratified with by 10 (so that the person would be motivated to promptly move away from the noxious stimulus to a more highly gratifying distance from it), and the more strongly (what would effectively be reflex) avoidance or withdrawal behavior would be excited by paradigm 11.

One might object that the nullification of all pain and other aversiveness would undermine the diagnostic skills of internists and surgeons. A rebuttal to this objection is entailed in the realization that LC-POA/MP recording devices, LC-POA/MP-based alarm systems, etc., could be used to monitor and record the time course of the intensity and anatomical distribution of POA- and MP-engendering pathophysiological processes (so that physicians would have precise and accurate records to refer to) and to inform physicians instantaneously of dangerously high intensities of these processes without the patient ever having to actually experience any of the POA or MP.

Pharmacologic and LCDBS approaches could readily be made therapeutically complementary to each other as is suggested by the observations that some pharmacologic agents facilitate self-(administered-brain-) stimulation behavior (26). Some LCDBS systems might even entail a pharmacologic component in the form, for example, of an implanted reservoir of some medication from which a minute quantum thereof would be released with each pulse or after every predetermined number, e.g., 10 or 100, of pulses of brain stimulation. It is clear that precautions against overdosing would have to be built into such prosthetic systems.

CONCLUSIONS

If, at times, in what will hopefully prove to be the relatively utopian future, we should want to nullify all selfishness, egocentricity, and loneliness and render all of us (or as many as would want to participate) optimally altruistic, that is, precisely as concerned about each other as about ourselves, we could accomplish this by telemetrically interconnecting every reward and punishment site in every participant's brain with its neuroanatomic counterpart in every other participant's brain. We would thereby affect <u>interindividual cerebral telemetric interconnectedness</u>, IICTI, by virtue of which we could all truly and thoroughly share our joys and sorrows (if there were still any of the latter despite LCDBS and possibly other aversiveness-annihilatory approaches) with each other.

Ideally, the neurobiological basis of the capability of experiencing distress of any kind, that is, possibly, for example, modes of functioning of certain intraneuronal structures in certain areas of the brain, might be prevented from ever developing as such by means of the individual's brain being subjected at one or more ontogenetically opportune times to certain drugs, electromagnetic or ultrasonic waves, recombinant repressor genes, antibodies, other immunological entities, or some combination thereof, directed against this neurobiological basis. Such prevention of development would be ideal because it would preclude even the potential for suffering.

The mind recognizes that there is a common denominator among the experiences of reading a book one enjoys, eating a food one enjoys, engaging in a favorite hobby or pastime, having sexual relations with a preferred partner, achieving a goal, etc. That common denominator, of course, is pleasure. The fact that the mind experiences pleasure as a distinct entity despite the great diversity among the numerous contexts and forms in which it can occur, suggests that there is an electrophysiological process common to all of these contexts and forms which could be objectively detected and quantified as an LC for pleasure (LC-PLE). This LC could be used as a measure of any and all of the various kinds of pleasure one can have to enjoy, such as happiness (which may be defined in the present context as pleasure which is consonant with one's idealism and/or aspirations), joy (which may be defined as pleasure which is anxiety-free and exhilarating), bodily or sensual pleasure, etc. Hence, LC-PLE, cumulated or averaged over time or in some other form, could be used to derive a concept and a value of a person's positive aspects of subjective state of being.

Analogously, one would expect that all distressing or <u>aversive</u> experiences could be quantified in terms of an LC-AVERS which could serve as a measure of a person's negative aspects of subjective state of being. The measure of a person's overall <u>subjective well-being</u> (SWB) would be a function of both LC-PLE, which, being a positive quantity, would add to its value, and LC-AVERS, which, being a negative quantity, would subtract from its value. The <u>standard deviation</u> of the distribution of all of the SWB scores (SD-SWB) of everyone in the world or universe could serve as an index of the equitableness of the distribution of happiness and other aspects of SWB among the members of the population of the world or universe and, hence, as an index of the degree of actualization of the democratic principle. Inasmuch as from an objective standpoint, each individual's SWB is equally as important as every other individual's SWB, the smaller the value of SD-SWB, the more equitable the distribution of SWB among the population.

Consistent with Bentham's ideal (27) of "the greatest happiness of the greatest number" the <u>quality</u> of our <u>world</u> or universe (QW) could be assessed in terms of the ratio or quotient of the magnitude of the sum <u>total</u> of the SWB scores of everyone in the world or universe (SWB-TOTAL: the larger its value, the better) divided by SD-SWB (the smaller its value, the better).

QW = <u>SWB-TOTAL</u>
SD-SWB

In a similar way as the Dow Jones Index provides a means of assessing broad-based economic strength, this ratio, the QW, would provide a means of assessing broad-based (ideally universal) happiness and other aspects of SWB. It might also serve as a means of determining whether or not the lot of humankind were actually improving over time, that is, whether or not the changes which will come about in the world will actually be constructive. The larger the value of QW, the more worthwhile, humanistic, and Heavenly we could consider our world to be.

References

1. Flor-Henry P., Yeudall L. T., Koles Z. J., Howarth B. G. Neuropsychological and power spectral EEG investigations of the obsessive-compulsive syndrome. *Biological Psychiatry* l4 (1), 119–130, February 1979.
2. Morihisa J. M., Duffy F. H., Wyatt R. J. Brain electrical activity mapping (BEAM) in schizophrenic patients. *Archives of General Psychiatry* 40: 719–728, July 1983.
3. Shagass C., Roemer R. A., Straumanis J. J. Relationships between psychiatric diagnosis and some quantitative EEG variables, *Archives of General Psychiatry* 39: 1423–1435, December 1982.
4. Stevens J. R., Livermore A. Telemetered EEG in schizophrenia: spectral analysis during abnormal behavioral episodes. *Journal of Neurology, Neurosurgery, and Psychiatry* 45 (5): 385–395, May 1982.
5. Stigsby B., Risberg J., Ingvar D. H. Electroencephalographic changes in the dominant hemisphere during memorizing and reasoning. *Electroencephalography and Clinical Neurophysiology* 42 (5): 665–675, May 1977.
6. Tucker D. M., Stenslie C. E., Roth R. S., Shearer S. L. Right frontal lobe activation and right hemisphere performance: decrement during a depressed mood. *Archives of General Psychiatry* 38 (2): 169–174, February 1981.
7. Wogan M., Moore S. F, Epro R., Harner R. N. EEG measures of alternative strategies used by subjects to solve block designs. *International Journal of Neuroscience* 12, 25–28, l981.
8. Gavriloy L. R. Use of focused ultrasound for stimulation of nerve structures. *Ultrasonics* 22: 132–138, May 1984.
9. Brown R. P, Kneeland B. Visual imaging in psychiatry. *Hospital and Community Psychiatry* 36 (5): 489–496, May 1985.
10. Mancini L. The prospect of a noninvasive brain stimulator. In progress.

11. Sem-Jacobsen C. W. Effects of electrical stimulation on the human brain. *Electroencephalography and Clinical Neurophysiology* 11: 379, 1959.

12. Heath R. G. Pleasure response of human subjects to direct stimulation of the brain: physiologic and psychodynamic considerations, pp. 219–243 in *The Role of Pleasure in Behavior* (R. G. Heath, ed.) Harper & Row, New York, 1964.

13. Moan C., Heath R. G. Septal stimulation for the initiation of heterosexual behavior in a homosexual male. *Journal of Behavioral Therapy and Experimental Psychiatry* 3 (1). 23–30, March 1972.

14. Higgins J. W., Mahl G. F., Delgado J. M. R., Hamlin H. Behavioral changes during intracerebral electrical stimulation. *A.M.A. Archives of Neurology and Psychiatry* 76: 399–419, 1956.

15. Holt L., Gray J. A. Septal driving of the hippocampal theta rhythm produces a long-term, proactive and non-associative increase in resistance to extinction. *Quarterly Journal of Experimental Psychology: Comparative & Physiological Psychology* 35B (z): 97–118, May 1983.

16. Heath R. G., Walker C. F. Correlation of deep and surface electroencephalograms with psychosis and hallucinations in schizophrenics: a report of two cases. *Biological Psychiatry* 20: 669–674, 1985.

17. White F. J., Wang R. Y. Differential effects of classical and atypical antipsychotic drugs on A9 and Al0 dopamine neurons. *Science* 22l: 1054–1057, September 1983.

18. Stein L., Belluzzi J. D., Ritter S., Wise C. D. Self-stimulation reward pathways: norepinephrine vs. dopamine. *Journal of Psychiatric Research* 11: 115–124, 1974.

19. Carpenter M. B. *Human Neuroanatomy* (7th ed) p. 589. Williams & Wilkins, Baltimore, 1976.

20. Heath R. G. Electrical self-stimulation of the brain in man. *American Journal of Psychiatry* 120 (6): 571–577, 1963.

21. Bishop M. P., Elder S. T., Heath R. G. Intracranial self-stimulation in man. *Science* 140 (whole no. 3565): 394–396, 1963.

22. Mancini L. How learning ability might be improved by brain stimulation. *Speculations in Science and Technology* 5 (1) (correspondence): 51–53, 1982.

23. Boivie J., Meyerson B. A. A correlative anatomical and clinical study of pain suppression by deep brain stimulation. *Pain* 13 (2): 113–126, June 1982.

24. Hosobuchi Y. Periaqueductal gray stimulation in humans produces analgesia accompanied by elevation of beta-endorphin and ACTH in ventricular CSF. *Modern Problems in Pharmacopsychiatry* 17: 109–122, 1981.

25. Plotkin R. Deep-brain stimulation for the treatment of intractable pain. *South African Journal of Surgery* 19 (14): 153–155, December 1980.

26. Jacques S. Brain stimulation and reward: "pleasure centers" after twenty-five years. *Neurosurgery* 5 (2): 277–283, 1979.

27. Rader M. *Ethics and the Human Community*, chapter 3, p. 91. Holt, Rinehart, & Winston, New York, 1964.

Medical Hypotheses
Medical Hypotheses (1992) 38, 350–351
©Longman Group UK Ltd 1992

Short Note
Ultrasonic Antidepressant Therapy Might Be More Effective Than Electroconvulsive Therapy (ECT) in Treating Severe Depression

L. S. MANCINI

ECT is widely acknowledged as an effective treatment for severe depression (1), but its drawbacks of entailing a seizure, general anesthesia, post-ictal confusion and memory disturbance can be problematical to the extent of preventing some patients from undergoing the treatments often enough to sustain relief. According to Higgins et al (2), stimulation of specific points on the surface of the temporal lobes reliably produces a pleasure response without causing a seizure, confusion or memory impairment. Due to reciprocal inhibition between the brain's reward and punishment pathways (3), such a response might be expected to relieve depression. Work done by Velling and Shklyaruk with animals, specifically rabbits (4), suggests that much stimulation could be produced in humans through the intact skull and scalp by using ultrasound focused from outside of the head. Ultrasound transducers contained in water-filled, bag-like stimulators, like those used by Gavrilov (5) to stimulate auditory nerves through intact skull, when placed in contact with the outside of the head, would eliminate the need for conductive gel, which can be messy. Since there would be no pain or other tactile sensation, anaesthesia would be unnecessary. Hence, externally-focused ultrasound could provide the same benefit as ECT, that is, relief of depression, without any of the drawbacks. And since the frequency of treatments would not be limited by these complicated factors, the therapeutic effects could be better retained over time and, therefore, more effective. Additionally, the basic approach of Velling and Shklyaruk could possibly be adapted for many other analgesic and therapeutic clinical applications such as treatment of pain, nausea and insomnia.

Date received: 8 November 1991
Date accepted: 10 February 1992

Acknowledgments
The author wishes to acknowledge with gratitude the encouragement and support of Drs. Robert A. Spangler, Michael R. Privitera, Nicholas J. LoCascio, C. Timothy Golumbeck, Louise A. Jameyson, Elizabeth Kelly-Fry, Robert J. Lanigan, Richard T. Mihran, Anthony A. Pace and Erica E. Wanecski.

References

1. Ahuja N. What's wrong with ECT? *American Journal of Psychiatry* 148, 5: 693–694, 1991.

2. Higgins J. W., Mahl G. F., Delgado J. M. R., Hamlin H. Behavioral changes during intracerebral electrical stimulation. *A.M.A. Archives of Neurology and Psychiatry* 76: 399–419, l956.

3. Stein L., Belluzzi J. D., Ritter S., Wise C. D. Self-stimulation reward pathways: norepinephrine vs. dopamine. *Journal of Psychiatric Research* 11: 115–124, 1974.

4. Velling V. A., Shklyaruk S. P. Modulation of the functional state of the brain with the aid of focused ultrasonic action. *Neuroscience and Behavioral Physiology* 18, 5: 369–375, 1988.

5. Gavrilov L. R. Use of focused ultrasound for stimulation of nerve structures. *Ultrasonics* 22, 3: 132–138,1984.

Medical Hypotheses
Medical Hypotheses (1992) 38, 349
©Longman Group UK Ltd 1992

Short Note
A Magnetic Choke-Saver Might Relieve Choking

L. S. MANCINI

When a person is choking on a piece of food and there is acute obstruction of the airway so that s(he) cannot breathe, the Heimlich maneuver or procedure can be effective in relieving the emergency (1). However, if someone were choking while dining alone or not in the company of anyone adequately skillful in the application of this maneuver, the problem might be solvable by means of a slight modification of a magnetic device (2) which is currently used in neurology, pneumonology and urology. It is used in these three fields, respectively, to assess disorder of the motor system, to measure diaphragmatic strength and to assess bladder and pelvic floor function. The device is a magnetic nerve stimulator which can stimulate deep or otherwise inaccessible nerves painlessly, contactlessly and without any need for removal of clothing. The stimulating coil is merely held in place either next to or a few millimeters away from the area of the body to be stimulated. With as little as perhaps 20% more power than that provided by currently available models, the stimulating coil could relieve choking by stunning the neuromuscular apparatus of the pharynx or larynx in such a way that it would momentarily relax and thereby release the food bolus and allow it to become reoriented so as to either be ejected, pass on down through the esophagus to the stomach or, in the worst case, down through the trachea into the lungs, in which case it would subsequently have to be removed in a clinical setting, perhaps by a nonsurgical, noninvasive technique involving suction. In any case, the emergency of being in danger of choking to death would be relieved as a result of application of the coil to the neck area. One way of demonstrating that the suggested method would be effective would be to test it on animals with obstructed airways whose temporal lobes were being pleasurably stimulated with ultrasound, as described in another paper (3), through the intact skull and scalp, so they would experience none of the distress or pain a human feels when s(he) accidentally chokes.

Date received 22 November 1991
Date accepted 10 February 1992

Acknowledgments

The author wishes so acknowledge with gratitude the encouragement and support of Drs. Robert A. Spangler, Nicholas J. LoCascio, Catherine Hall LoCascio, Michael R. Privitera, Erica E. Wanecski, Anthony V. Ambrose, Kenyon A. Riches, William and Denise Appel and his family.

References

1. Heimlich H. J. A life-saving maneuver to prevent food-choking. *JAMA* 234: 398, 1975.
2. Barker A. T., Freeston I. L., Jalinous R., Jarratt J. A. Magnetic stimulation of the human brain and peripheral nervous system: an introduction and the results of an initial clinical evaluation. N*eurosurgery* 20, 1: 100–109, 1987.
3. Mancini L. S. Ultrasonic antidepressant therapy would be more effective than electroconvulsive therapy (ECT) in treating severe depression. *Medical Hypotheses* 38: 350–351.

Medical Hypotheses
Medical Hypotheses (1992) 31, 201–207
©Longman Group UK Ltd 1990

Riley-Day Syndrome, Brain Stimulation and the
Genetic Engineering of a World Without Pain

L. S. MANCINI

Abstract — Riley-Day Syndrome, a genetic disorder in which there is impaired ability or inability to feel pain, hot and cold, is cited as an example of evidence that the commonplace notion that life cannot be painless is not necessarily valid. A hypothesis is presented to the effect that everything adaptive which is achievable with a mind capable of experiencing varying degrees of both pleasure and pain (the human condition as we know it) could be achieved with a mind capable of experiencing only varying degrees of pleasure. Two possible approaches whereby the human mind could be rendered painless are a schematically-outlined genetic approach which would or will probably take thousands of years to implement, and a brain stimulation approach that could be effected by means of a noninvasive contactless, transcranial, deep neuroanatomic-site-focusable, electromagnetic and/or ultrasonic (and/or, conceivably, other kind of) brain pacemaker which could be developed within a few years. In order to expedite the relief of all kinds of suffering and the improvement of the human condition in general, it is advocated that prompt and concerted research effort be directed toward the development of such a brain pacemaker.

Introduction

In this article the concept of pain should be construed in its broadest sense. It should be thought of not merely as bodily suffering but rather as any unpleasant or distressing experience, whether it be of bodily pain, nausea, dyspnea, pruritus, hunger, thirst, fear (anxiety), depression, anger, etc. Most people seem to regard pain as a necessary evil of life. Such rationalizations as "no pain, no gain," which are widely invoked, may be adaptive in terms of helping one to deal with life's current reality, but this does not mean they are or will prove to be insurmountably valid for all time. On the contrary, one will find that the people who are the most productive and efficient at their work tend to be the ones who enjoy it the most.

Or people will ask, "How would you be able to recognize pleasure if you never experienced pain?" On the contrary, one does not need to have been tortured in order to enjoy fine music or fine dining. Or people will contend that happiness is a matter of free will. In effect, those who are happy are so because they choose and work to be. Those who are unhappy are so because they do not choose or work to be happy. However, in view of the now well-documented biological and, in some cases, hereditary bases of many of the depressive, anxiety, and psychotic disorders, to say nothing of adverse external circumstances such as the occurrence of natural disasters, over which one cannot possibly (except, conceivably, superstitiously) have any control, any self-deterministic theory of happiness level seems seriously flawed or altogether incredible.

Or people will rationalize: "If you were happy all the time, then you would be bored." On the contrary, this statement makes no sense because happiness and boredom are mutually exclusive states. If one were to consider only the happiest (the most enthusiastic and contented) one or two percent of the human population, it would be surprising if one would find that these people spend even a small fraction of their time being bored. Hence, even given the existing state of affairs, it is not necessarily humanly impossible to be happy most of the time. Some electrophysiological ways in which pleasure and pain could be measured and quantified are suggested below.

Riley-Day Syndrome

A rare genetic disorder known as Riley-Day Syndrome or familial dysautonomia, i.e., FD (1, 2, 3, 4), which has an autosomal recessive means of transmission, was first identified in 1949. It occurs almost exclusively in descendants of the Eastern European Ashkenazy branch of Judaism, although it has been diagnosed occasionally in members of other religious, ethnic, and/or racial groups.

The syndrome is evident from the time of birth in terms of difficulty in feeding, episodes of unexplained fever and pneumonia, and failure to thrive. Its symptoms and signs as a lifelong disorder include defective lacrimation with an inability to shed tears when crying, corneal ulceration, absent corneal, axonal, and tendon reflexes, unstable blood pressure with episodes of hypertension and postural hypotension, unstable body temperature, vomiting spasms, profuse sweating, sialorrhea, impairment of vestibular function, repeated infections, an initial delay in mental development (with the subsequent achievement of intellectual parity with one's peers by age 4), difficulty or inability to suck or chew with impaired pharyngeal and esophageal motility, esophageal and intestinal dilation, absence of taste buds or fungiform papillae with inability to taste food, a marked tendency to develop symmetric blotchy erthematous skin rashes, especially in connection with eating or emotional stress, emotional lability with a nervous system which is unstable in the sense that strong emotions, whether pleasing or distressing, frequently lead to episodes of loss of consciousness, stunted growth and, most significantly from the standpoint of this article, an impaired ability or *inability* to feel pain, and an impaired ability or inability to feel (or distinguish between) hot and cold, although there is relative preservation of pressure and tactile sense.

The syndrome is considered to represent a disturbance of both sensory and autonomic functions, both parasympathetic and sympathetic. Compared to individuals who are not afflicted with the syndrome i.e., neurologically normal individuals), there is a diminution in the number of sympathetic and parasympathetic ganglion cells and, to a lesser degree, in the number of nerve cells in the sensory ganglia. There is a paucity of small myelinated and unmyelinated nerve fibers, which explains the impairment of temperature and pain sensation. There is increased excretion of homovanillic acid and decreased excretion of vanillylmandelic acid and methoxyhydroxyphenylglycol. There is also an abnormally low concentration of serum dopamine beta-hydroxylase, the enzyme that converts dopamine to norepinephrine.

There are about 300 known cases in the United States, but the true incidence, which would amount to a larger number, among American Jews, is estimated to lie between 1 in 10,000 and 1 in 20,000 with a carrier frequency of 1 in 50 to 1 in 70. According to Bundey and Brett (1), 25% of afflicted children are dead by age 10 and 50% by age 20, usually as a result of pulmonary problems secondary to bronchial hypersecretion or inhalation of stomach contents during attacks of vomiting. Apparently due to consciousness raising brought about by the New York–based Dysautonomia Foundation and the concerted efforts of a pediatrician, Dr. Felicia Axelrod (2), who has centered her life's work on FD, improved supportive and symptomatic treatment has resulted in more afflicted children surviving to adulthood. Nonetheless, so far no one known to have this condition has survived beyond the fifth decade of life. There is, as yet, no definitive treatment.

Although FD probably originated well before its date of initial identification (1949) it is at least ironic that a genetic condition involving an inability to experience bodily pain would emerge primarily among a group of people who had been subjected, unfortunately and unpardonably (over a period of centuries or millennia), to extraordinary amounts of pain (which culminated) only a few years earlier, during the heinous period of Nazi domination connected with World War II (2, 3, 5).

The underlying implications of Riley-Day Syndrome

Although from a superficial standpoint the existence of FD as a disease entity may seem to underscore the indispensability of pain to the processes whereby an organism adapts to its environment, closer scrutiny will reveal that this indispensability may be more illusory and circumstantial than it is real and immutable. The most important point to be deduced from FD is that life without a phenomenon which most of us assume is an unavoidable part of life (that is, the capability of experiencing bodily pain) is not only conceivable but actually occurs in some cases. If FD did not exist, it would be easier for anyone to claim that life without the capability of experiencing bodily pain is an impossibility.

If the genetic modification which underlies FD can result in a life without bodily pain, then it is quite conceivable that other genetic modifications could result in lives devoid of the capabilities of experiencing any and all other forms of pain, including everything from nausea to frustration and hostility (variants or subtypes of anger).

In fact, there is another condition, even rarer than FD, known as congenital indifference to pain (1), in which the individual does recognize a painful stimulus as such but perceives it as no more distressing than a touch or a tickle, that is, not distressing at all. These patients can learn to take precautions that will minimize the possibility of trauma. So far, the nervous systems of such individuals have not been shown to be abnormal in any way (1). Nonetheless, this condition of lifelong unreactivity to pain does appear to adhere to a genetic pattern of inheritance, in some cases autosomal dominant, in others autosomal recessive. Because it appears to be possible to compensate adaptively for this condition, it would appear to be a better candidate for genetic modeling for a painless world than FD, but it too leaves much to be desired, both because it tends to be maladaptive and because it does not preclude forms of pain other than the distress of bodily pain.

How life could be painless without being maladaptive

Let us now broach the question of how, in psychodynamic terms, we could go through life as adaptively (or more so, and certainly more enjoyably) without the capability of experiencing any kind of pain as we do in our current condition of being equipped with that capability. For example, if we did not have anxiety—specifically, the fear of getting killed—what would prevent us from driving our cars into the oncoming traffic? An answer is as follows. We would not want to genetically engineer ourselves so that we would be rid of the anxious impulses that deter our driving into traffic without putting anything adaptive in their place. However, what could adaptively be put in place of these anxious impulses would be pleasure-diminishing impulses with or without the added adaptive benefit of reflexly avoidant impulses. Hence, rather than producing pain (anxiety), the thought or anticipation of the possibility of driving into oncoming traffic would produce a marked diminution of the high level of baseline pleasure, with or without an accompanying unconsciously motivated reflexive avoidance response, which would prevent the injurious behavior just as reliably as the anxiety currently does. The option of having unconscious reflex avoidance associated with any perception of potential danger, while not strictly necessary in terms of protective motivational wherewithal, might facilitate faster reaction times (because less interneuronal processing would be entailed) than would pleasure diminution alone associated with any perception of danger.

Furthermore, a proportional factor could be built into the system (the genetically engineered mind) so that the more dangerous the situation (for example, the closer one's actual position to the oncoming traffic), the more marked the diminution of pleasurable impulses (and, possibly, the stronger the unconsciously motivated avoidance responses) would be.

By generalizing from this example, it is possible to appreciate that everything adaptive which can be achieved with a mind capable of experiencing varying degrees of both pleasure and pain (the current human condition) could be achieved with a mind capable of experiencing only varying degrees of pleasure or a mind capable of experiencing such pleasure simultaneously with varying intensities of unconsciously mediated avoidance behavior. The possibility of leading a strictly painless yet adaptive life would constitute an improved human condition.

This improved condition could be achieved in either one of two different ways, with each of them having its respective pros and cons. One way in which it could possibly be achieved would be with a contactless, noninvasive, transcranial brain pacemaker, which could also be called a noninvasive neuroprosthesis or brain stimulator. The essence of the method whereby brain pacemaking or brain stimulation could be used to effect the improved human condition is delineated in another paper (6) and, therefore, need not be repeated here.

Brain stimulation experiments done on animals confirm the widely acknowledged point that "stress kills," which is to say that the unpleasant aspect of stress or distress affects organisms adversely not only in terms of the quality of life, but also in terms of the duration of life. Prolonged, unconditional, unavoidable stimulation, lasting 24 hours or more, of areas within an animal's brain which cause the animal to show all of the signs of extreme distress (i.e., "punishment centers," areas to which an animal will promptly terminate stimulation if given the means to do so) has been observed to actually cause the animal to become severely ill and die (7). Perhaps heedless to say, such experiments are horrendously cruel and, while carrying out such experiments even once is unconscionable regardless of the extent to which scientific knowledge might be augmented thereby, certainly should never be done again.

How genetic engineering could be used to effect a painless improved human condition

There is currently a good deal of speculation and controversy regarding the prospect of mapping and sequencing the entire human genome. Dr. Leroy Hood, a pioneer in the field of biotechnology (8), estimates that "it will take us at least hundreds of years to decipher the multitude of messages contained in the human genome."

We know that some people have a low threshold of pain and other people have a high threshold of pain. And this would seem to be true regardless of what kind of pain we are talking about. It is also quite likely, as is exemplified by FD, that there is a strong genetic basis for the height of a person's threshold for any kind of pain. Let us consider two different kinds of people. The first kind of people become greatly distressed when confronted with adverse circumstances and only tolerably placated when confronted with favorable circumstances. Let us describe these people as *pain-dominated* (PAD). The other kind of people become only mildly displeased when confronted with adverse circumstances and become greatly elated when confronted with favorable circumstances. Let us describe these people as *pleasure dominated* (PLD).

The majority of most present-day populations would fall somewhere in between the two extremes of being PAD and PLD. However, let us suppose that for any given person there is some electrophysiologically discernible characteristic (of the computerized-spectral-analyzed-electroencephalogram or EEG, magnetoencephalogram or MEG, electromyogram or EMG, or whatever) which can be detected whenever the person is experiencing any kind of pleasure. This characteristic can be referred to as the pleasure characteristic (PLC). Because the mind gives the same "pleasurable" label to all different kinds of pleasure (in other words, the mind is aware of both the differences and the fundamental *similarity* among the various different kinds of pleasure), it makes sense that there would be some potentially electrophysiologically detectable, common denominator of all of them. Similarly, suppose that for any given person there is some electrophysiologically discernible characteristic which can be detected whenever the person is experiencing any kind of pain. This can be called the pain characteristic (PAC).

Then, by measuring any person's amplitudes, duration, and frequency of occurrence of PLC and PAC, one could determine whether the person is more PLD or more PAD. If we were to measure and record the PLC and PAC values for every person in a population over a period of time, we would then be able to identify both those who arc extremely PLD and those who are extremely PAD. Suppose the 1% of the population who are most PLD (1%-PLD) and the 1% who are most PAD (1%-PAD) have been isolated. One would probably then be able to verify that the 1%-PLD and the 1%-PAD groups, respectively, (approximately at least after individual differences in stress exposure have been taken into account) constitute or at least correlate positively with the 1% of the population with the highest and lowest pain thresholds. Using some arbitrary scale of pain threshold, suppose the average pain threshold value of the 1%-PLD group is 100 and the corresponding value for the 1%-PAD group is 10. After the task of having sequenced the entire human genome has been completed, suppose we move on to the task of sequencing each person's entire genome or, at least for the purpose in question here, the task of sequencing the genome of each person in either of these two 1% subpopulations.

One should then look for systematic differences between the DNA sequences of the 1%-PLD group and the 1%-PAD group. The likelihood seems high that there would be at least one common denominator, or possibly a number of different patterns of common denominator(s), of DNA sequencing among the 1%-PLD group (PLD-DNA) and at least one common denominator, (different from PLD-DNA) of DNA sequencing among the 1%-PAD group (PAD-DNA). Then, by noting the trend of differences between the DNA-sequence common denominator patterns for the pleasure-dominated group and the DNA-sequence common denominator patterns for the pain-dominated group, one could possibly extrapolate a DNA sequence common denominator that would characterize people with an ultra-high pain threshold of, for example 1,000. Let us refer to this extrapolated sequence as a super-pleasure-dominated DNA or super PLD-DNA sequence. Anyone with such a DNA sequence in her or his genome would have a very high (possibly higher than any level or value which nature, unassisted, has bestowed upon anyone) threshold of pain, perhaps comparably as high for all kinds of pain as the thresholds for mechanical and thermal pain of the prototypical FD sufferer, which could mean unreachably high for all kinds of pain.

Any individual with the extrapolated DNA sequence in his or her genome might actually still be mildly pleased (or have a neutral affect) when confronted with adverse circumstances and would undoubtedly be joyously elated when confronted with favorable circumstances. Such an individual could never experience any pain. Her/his mood could only vary between neutral affect when confronted with catastrophic circumstances (but the pleasure-seeking nature of all organisms, regardless of height of pain threshold, would still motivate such a person to do whatever possible to calmly undo or compensate for catastrophic circumstances), and joyous elation when confronted with even the ordinary homeostasis conducive circumstances, such as the pervasiveness of ample amounts of oxygen for the purpose of breathing, which most of us take for granted.

Then, one could implement in vitro genetic engineering by microinjecting every fertilized human egg with the recombinant super PLD-DNA sequence(s) that would replace the existing homologous (mediocre) PLD-DNA and/or PAD-DNA sequence(s) so that all humans born after the perfection of the recombinant super PLD-DNA sequence microinjection technique would have a pain threshold of approximately 1,000 and be unable to experience pain of any kind.

Or, when and if in vivo genetic transformation techniques are ever perfected, whereby every cell in the body of a living organism at any stage in its lifespan can simultaneously undergo homologous insertion of any desired recombinant DNA sequence encoding for any desired gene(s) coupled with the deletion of the homologous, less desirable DNA sequence encoding for any less desirable gene(s), then all individuals alive at the time of the achievement of such technical capability, regardless of their age at that time, could be converted from being PAD- or PLD-type people to being super-pleasure-dominated people incapable of experiencing pain of any kind.

But what if these elation-prone, super-pleasure-dominated, genetic transformers with unreachably high pain thresholds prove to be aimlessly and recklessly euphoric, like many present-day psychiatric patients who are afflicted with mania, as in manic-depressive illness? Such individuals would have little or no inclination toward concerted, constructive, and productive activities (i.e., learning and working; work that is of value to society), and a marked inclination toward consumptive or libidinal activities such as wild spending sprees, gambling, alcohol abuse, sexual indiscretion, overeating, etc. Even though they would be incapable of suffering, they would be relegated to relatively unfulfilling, hence dull, minimally pleasurable lives, because nothing and no one would reward them for their unproductivity. Their lives would be devoid of a sense of purpose which is essential to happiness. They would be a burden to society and to themselves. Their problems could be solved by implementing methods whereby their vast potential pleasure could be actualized and harnessed or channeled into constructive, productive, adaptive activities and pursuits.

One way of doing this would be by implementing the brain stimulation paradigms delineated in another paper (6), whereby learning and working could be made at least as pleasurable and probably (by dint of the intrinsic pleasurableness of diversification of knowledge and skills) more pleasurable than consumptive or libidinal activities.

Another way, a genetic way, of accomplishing this would be by looking for and isolating common denominators of DNA sequencing that characteristically are present within the genomes of individuals (regardless of whether they are PLD or PAD) who do function productively and adaptively. And then, by in vivo insertion of the productivity-/adaptiveness-characteristic DNA sequences into every cell in the bodies of these unproductive, maladaptive super PLD individuals, which would effect homologous replacement of DNA sequences encoding for unproductive, maladaptive tendencies with DNA sequences encoding for productive, adaptive tendencies, they could convert themselves from pleasureless, painless, unproductive members of society into elated, productive members thereof. Hence, the goal of adaptive, painless, and pleasurable living could be achieved genetically as well as by brain stimulation.

The pros and cons of the two different approaches

The genetic engineering approach to achieving painless living would be the definitive one, far superior to brain pacemaking, primarily because, with the former, the individual would not have to be dependent upon or encumbered with external gadgetry or subject to conceivable side effects of long-term brain stimulation. However, the drawback of the genetic approach is that it may take a very long time, perhaps thousands of years, to implement. Such amounts of time would certainly make sense on the time scale of Dr. Hood. In sharp contrast to this greatly extended time scale, noninvasive, contactless brain stimulation or pacemaking, which could be used to accomplish essentially the same goal, that of painless yet adaptive living, could be developed within a few years. Dr. Robert G. Heath, a pioneer in the use of surgically implanted electrodes to effect neuropsychiatrically relevant brain stimulation, has indicated that an ultrasound-emitting device could be built (ostensibly as early as any time between the present moment and the early part of the 21st century) which could activate the brain's "pleasure centers" without having to go inside the skull. And, in line with his claim is a prediction that, by the year 2005, family physicians will be using such a device on a routine therapeutic basis (9).

All of the technological ingredients that would have to be brought together in order to construct such a device already appear to exist. The combined use of electromagnetism and ultrasound (as opposed to ultrasound alone), as suggested by W. J. Fry (10) and affirmed by his brother, F. J. Fry (11), might more readily facilitate the goal of developing a contactless, noninvasive brain pacemaker capable of exciting or suppressing any small or large area(s) in the living human brain. The principal drawbacks of brain pacemaking would be that it would entail the possibility of periodic equipment failure or malfunctioning, the burden (even if a very small, lightweight one) of having to carry around a pacemaker wherever one wanted to go while still having the benefit of its use, and the possibility of some as yet undetermined side effect(s) which, according to the observations of Barker et al (12), are unlikely to prove prohibitive.

Genetic engineering could enable each person to be whatever s(he) wants whenever s(he) wants

Dr. Hood also predicts (13) that "It isn't that [through genetic engineering] we'll be able to design individuals whose intelligence is increased by a factor of three. It isn't that we'll be able to change physical attractiveness or emotional stability." Given the virtually universal human motive toward self-improvement, it appears doubtful that this prediction will prevail through the extent of time. Let us not think in the authoritarian terms of some individuals genetically engineering the characteristics of others. Instead, let us think in the egalitarian terms of each individual genetically re-engineering herself/himself according as s(he) pleases. What is being suggested here is that in the distant future, by means of in vivo genetic transformation techniques effected with recombinant DNA or some other biotechnological tool(s), it will be possible for any person (or other kind of organism) to be an introverted, academically-oriented, purple-haired, orange-eyed, 10-foot-tall white male with an IQ of 160 on any given day and a party-going, humorous, green-haired, green-eyed, three-foot-tall green female with an IQ of 200 on the next day. Stated in more general terms, it will become possible for each one of us (that is, anyone alive during the future era in question) to be whatever we want to be whenever we want to be. Some of us may choose to take a DNA pill that will cause us to sprout a pair of wings whenever an automobile is not readily available to us. Granted: all of this may be thousands of years away, but compared to the eternity that stretches ahead, the amount of time in question is minuscule.

Some may object that if each of us were able to change our mental and/or physical characteristics at any given time, then the notion of individual identity would be essentially lost. Undoubtedly, the objection is at least partly valid (however, the philosophical implications will not be delved into here), but the advantages of such a greatly augmented arena of potential endeavor would seem to greatly outweigh the disadvantages. In a world equipped with virtually instantaneous, or at least high-speed, self-determined in vivo genetic transformation, no one would ever have any reason to feel inferior to or less fortunate than anyone else, because whatever characteristic(s) one might envy in another person, one could incorporate into one's own being almost as quickly as the envious impulses could emerge. Hence, the ideal of all humans being equal could be realized in terms of each individual having the same (infinitely variable) genetic potential. If this were the case, both egotism and the competitive spirit would become extinct, but their disappearance from the world really would not be a substantial loss; in fact, it would be a gain for everyone.

Conclusion

In the meantime, that is, the intervening thousands of years between now and the successful implementation of genetic engineering techniques which might create the possibility of painless living, it would be advisable to develop a brain pacemaker and then to determine whether or not its drawbacks could be eliminated or minimized to a tolerable level. Moreover, as one of its virtually limitless potential therapeutic applications, it is conceivable that such a pacemaker, by suppressing maladaptive neuronal impulses (such as those underlying the vomiting attacks) and by exciting adaptive impulses (such as those that could underlie better coordinated pharyngeal and esophageal motility during eating) could provide more effective therapy for Riley-Day Syndrome than any that is currently available by minimizing or eliminating such problems as the aspiration of vomitus, impaired eating ability, etc.

References

1. Bundey S., Brett E. M. *Genetics and Neurology*, p. 206. Churchill Livingstone, Edinburgh, 1985.
2. Axelrod F. B., Sein M. E. *Caring for the Child with Familial Dysautonomia – a Treatment Manual*. Dysautonomia Foundation Inc., New York, 1982.
3. Adams R. D., Victor M. *Principles of Neurology*, 3rd edition, p. 991. McGraw-Hill, New York, 1985.
4. Axelrod F. B. *Report on the Dysautonomia Research Symposium*. Dysautonomia Foundation Inc., New York, May 1987.
5. Chusid J. G. *Correlative Neuroanatomy and Functional Neurology*, 16th edition, p. 148. Lange Medical Publications, Los Altos, California, 1976.
6. Mancini L. Brain stimulation to treat mental illness and enhance human learning, creativity, performance, altruism, and defenses against suffering. *Medical Hypotheses* 21: 209–219, 1986.
7. Guyton A. C. *Textbook of Medical Physiology*, 7th edition, p. 680. W. B. Saunders Company, Philadelphia, 1986.
8. Hood L. Biotechnology and medicine of the future. *The Journal of the American Medical Association*. March 25: 1837–1844, 1988.
9. Clarke A. C. *July 20, 2019, Life in the 21st Century*, p. 223. Macmillan Publishing Company, New York, 1986.
10. Fry W. J. Electrical stimulation of brain localized without probes – theoretical analysis of a proposed method. *The Journal of the Acoustical Society of America* 44, 4: 919–931, 1968.
11. Fry F. J. Personal communication, May 27, 1987.
12. Barker A. T., Freeston I. L., Jalinous R., Jarratt J. A. Magnetic stimulation of the human brain and peripheral nervous system: an introduction and the results of an initial clinical evaluation. *Neurosurgery* 20. 1: 100–109, 1987.
13. Davis J. Leroy Hood: automated genetic profiles. *Omni*, November: 116–156, 1987.

REPRINTED FROM

Speculations in

Science and

Technology

Speculations in Science and Technology, Vol. 16, No. 1, page 78.

Letter to the Editor

A proposed method of pleasure-inducing biofeedback using ultrasound stimulation of brain structures to enhance selected EEG states

Lewis S. Mancini

A noninvasive method of stimulating human brain structures to induce pleasure involving focused ultrasound waves is proposed. Such a device in combination with an EEG analysed for desirable mental states, could be used to enhance thinking, ameliorate neuroses and treat chronic pain.

The seeking of pleasure and the avoidance of pain are perhaps the root behavioral mechanisms in humans. It is proposed that a device which analyzes an EEG for specific patterns and stimulates, through a noninvasive technique, the pleasure centers of the brain when a desired pattern is achieved, would be useful in enhancing thought, overcoming neuroses and in treating chronic pain.

The novel mechanism in such a device would be an ultrasound beam focused from outside of the skull onto a key area of the subject's brain. Research in the former Soviet Union by Velling and Shklyaruk[1] has shown that a focused ultrasound beam is capable of selectively stimulating neuron activity in localized areas of animal brains. Velling and Shklyaruk found that ultrasonic waves in the intensity of 1–100 mW cm^{-2} and of a duration of more than one second led to increased bioelectrical activity in the brains of cats and rabbits. The effects were reversible; i.e., no permanent damage was detected. It is postulated that the mechanism of the ultrasound's action may be a change in the permeability of neuron membranes through elastic deformation of the membranes.

A hypothetical ultrasound setup could be adapted to stimulate the septal area of the human brain, inducing a pleasurable response and an increased level of consciousness. Heath[2] reported that electrical stimulation of electrodes implanted directly in the brains of 54 schizophrenic and epileptic patients led to pleasurable responses in most patients when the septal area was stimulated. Other areas inducing positive feelings were the medial forebrain bundle and the interpenduncular nuclei of the mesencephalic tegmentum. "With septal stimulation the patients brightened, looked more alert, and seemed to be more attentive to their environment ..." Heath reported.

Heath also found "… (s)triking and immediate relief from intractable physical pain" when the septal region was stimulated in six patients with advanced cancer. This result suggests that the proposed ultrasound device might be useful in the management of chronic pain. Heath also speculated that the stimulation of the septal region might help people with neuroses overcome patterns of experiencing "emergency" emotions during normal situations.

Ultrasonic stimulation of the septal region would not necessarily be merely a new avenue into meaningless hedonistic play without a serious purpose attached. The author suggests that an electroencephalograph could monitor the subject's mental state during a task or activity, and a computer analyzing the tracings could activate the ultrasound when a desired mental state is achieved. Research has shown correlations between EEG patterns and various mental states, including alertness, concentration and memory retrieval.[3] Such correlations, though, have not been determined to be definitely reliable. Thus, the proposed biofeedback device probably would have to be calibrated for each individual. The device, used in such a manner, would be a powerful tool for enhancing an individual's skills in thinking and learning.

A previous line of research begun in the 1950s and continued in the 1960s involving direct stimulation by electrodes of human subjects' brains has not been followed. Presumably, this is due to two factors: (1) a moral, puritanical objection to such a short-cut to pleasure; and (2) the disturbing prospect of having electrodes implanted into one's brain. It is beyond the scope of this paper to examine the moral problem of pleasure as a phenomenon. However, Heath[2] found that electrical stimulation of pleasure centers in the brain, though perhaps creating a risk of dependence, was not addictive. The second shortcoming of previous research, the implantation of electrodes, would be overcome by the proposed mechanism, as ultrasound waves involve no surgical penetration of the brain tissue.

The potential uses of a noninvasive device for evoking pleasure and erasing pain in a human subject are numerous. It is suggested that research into this thought-provoking area would have significant benefits in the fields of education, psychology and medical pain management.

References

1. Velling V. A. and Shklyaruk S. P. 1988 Modulation of the functional state of the brain with the aid of focused ultrasonic action. *Neuroscience and Behavioral Psychology*, 18 (5), pp. 369–375.
2. Heath, R. G. 1964. Pleasure response of human subjects to direct stimulation of the brain: physiologic and psychodynamic considerations. In: Heath R. G. (ed.) *The Role of Pleasure in Behavior*, pp. 219–243, Harper and Row, New York.
3. Low, M. D. 1987. Psychology, psychophysiology, and the EEG. In: Niedermeyer, E. and da Silva, F. L. (eds), *Electroencephalography. Basic Principles, Clinical Applications and Related Fields*, pp. 541–548, Urban and Schwarzenberg.

Received: 20 May 1992; accepted: 8 June 1992

MENTAL HEALTH WORLD

WAITING HOPEFULLY

by Lewis S. Mancini

From the time of earliest recollection, both my mother and physician father expressed definite expectations that I should become a physician. Early on, it became obvious to everyone involved that it takes me at least twice as much time as most people to learn any subject matter or accomplish any task. The problem was incomprehensible not only to my parents and teachers, but also to me.

During my junior and senior years of college, I became significantly depressed. The prospect of disappointing my parents as to career choice seemed unconscionable. On the other hand, the necessary brain power for a medical career simply didn't seem available to me. Suddenly, one day during July 1972, as I was dozing off in organic chemistry class, a cogent and somewhat reassuring understanding of the basic problem popped into awareness; the reason for the abnormal slowness is twofold:

1. An *attention deficit* associated with excessive daydreaming and rumination;
2. An abnormally high level of (learning- and work-related) *performance anxiety*, associated with an extreme fear of failure and fear of making mistakes.

Hence, due to deficient attention and intense anxiety, concentration becomes markedly impaired and progress toward goals becomes obstructed.

As soon as the optimistic notion of finding a therapeutic way of simultaneously enhancing attention and diminishing anxiety began to circulate through my mind, the depression began to diminish appreciably. This notion or prospect constituted a bright light at the outlet of a depressive tunnel, a clear sparkle of hope.

Although it was fairly apparent that professional help might entail the most effective possible pharmacologic or behavior-modification therapy, at the time there was neither sufficient financial wherewithal nor security of self-esteem to take the necessary steps to obtain such help. Consequently, the only readily accessible (admittedly less than optimal) treatment method seemed to be a regimen of increased daily caffeine intake (increased from one to five or six cups a day).

By virtue of the fact that the megadoses of caffeine improved the attention deficit (by enhancing alertness) much more markedly than they worsened the anxiety, a significant net benefit was acquired. This benefit, while not dramatic, was enough to facilitate passing grades in all premedical coursework.

Having mediocre grades, of course, I didn't gain admission to a U.S. medical school, but was fortunate to receive an acceptance from a foreign medical school. In the context of psychiatry residency (initiated in '83), my day's work typically stretched from 8:30 or 9:00 a.m. till midnight or beyond. I was getting very burned out, very quickly. My bosses noticed this. And in '85 they laid me off with an indeterminate return date.

I went on to total psychiatric disability with a diagnosis of obsessive-compulsive disorder (OCD), substantiated by various symptoms, including many hours spent each day praying and engaging in superstitious rituals of many different kinds, intended to ward off evil influences.

Time-wasting rituals included excessively meticulous and repetitious grooming (five to six hours to get ready to leave the house, including a full hour in the shower to ensure "adequate" cleanliness). Adherence to ideas that certain colors, numbers and positions of objects in the physical environment were either good or evil, resulted in arranging and, almost incessantly, rearranging them, in my mind or in space in highly idiosyncratic ways to maximize superstitious advantages and minimize superstitious disadvantages. Another major problem was checking and rechecking, over and over again, to ascertain that no errors were made in context of any tasks undertaken.

In 1985 an electroencephalogram (EEG or brain wave test) demonstrated abnormal neurological function, in particular, abnormal, slow activity on the left side of the brain. Every plausibly appropriate medication and virtually every conceivable kind of behavior modification were determined (by the psychiatrist and two clinical psychologists) to be ineffectual. So, from a practical standpoint, what the OCD diagnosis and abnormal EEG seemed to signify was both a form of attention deficit disorder (ADD) and learning disability (LD). These inferences were confirmed in '92 by a psychologist who administered a comprehensive battery of psychometric tests.

At this point, I'm hopefully and patiently awaiting the emergence of a "wonder" drug that'll simultaneously enhance alertness and attention, while diminishing task-related performance anxiety. Another possibility might be some therapeutic adaptation and application of the kinds of physical phenomena, such as ultrasound and electromagnetic forces, which are in widespread use for medical imaging and diagnosis.

If they can be used noninvasively (that is, nonsurgically) to visualize and analyze minute details of structure of anatomic areas of medical interest, then isn't it conceivable they might also be used to therapeutically and noninvasively influence their *functions*? This is a question which can only be answered by way of extended amounts of time and considerable outlays of effort. In the meantime, I'm waiting patiently and doing the best I can.

Curriculum Vitae

LEWIS S. MANCINI

EDUCATIONAL BACKGROUND

B.S.   -   1973 – Psychology Major, Trinity College (Connecticut), 1969–73
- Coursework in Master's of Natural Sciences Program at State University of New York at Buffalo (S.U.N.Y.A.B.), 1973–74
- Coursework in undergraduate electrical engineering at the University of Utah, 1975–76
- Coursework in Bioengineering Master's of Engineering (M.E.) program at the University of Utah, 1977–78

M.D.   -   1983 – St. George's University School of Medicine (S.G.U.S.O.M.), Grenada, West Indies, 1978–83

Certificate of Psychiatric Residency - 1985 – S.U.N.Y.A.B. Affiliated Hospitals Psychiatric Residency Training Program, 1983–85

A.A.S.   -   1988 – (with high distinction) Associate in Applied Science in Electroencephalography (EEG) Technology from Niagara County Community College (N.C.C.C.) (program cosponsored by S.U.N.Y.A.B. School of Medicine), 1986–88
- Independent study in genetics (Life Sciences Department) at N.C.C.C. under Dr. Nicholas LoCascio, 1988
- Tutorial in Biophysics at S.U.N.Y.A.B. under Dr. Robert Spangler, 1989
- Writing class with editor Denise Sterrs and editor/novelist William Appel (author of *Whisper … he Might Hear You*, *The Watcher Within*, etc.), 1991
- Independent studies in Biophysics at S.U.N.Y.A.B. under Dr. Robert Spangler, 1992

PROFESSIONAL BACKGROUND

1994   Visited Drs. Scott Lukas of Harvard, Richard Pavelle, Zeb Hed of Massachusetts Institute of Technology (MIT), Mr. Dick & Mrs. Eleanor Grace of Brain Research Institute on Cape Cod, and Dr. Conan Kornetsky of Boston University

1993   Guest speaker at seminar at the Pennsylvania State University, Department of Engineering Science and Mechanics, topic: "Noninvasive Brain Pacemaker or a Real-Life Thinking Cap" (2-15 to 2-17)

1992   Visited Drs. Joie P. Jones and Patricia C. Rinaldi, Departments of Radiology, Surgery and Neurosurgery, School of Medicine, University of California at Irvine, to observe experiments in ultrasonic brain stimulation (6-22 to 6-27)

1989   Guest speaker at EEG technology lecture, Erie County Medical Center, topic: "Psychiatry and the EEG" (December)

1988   Visited Dr. Reza Jalinous at Massachusetts General Hospital and Dr. Vern Gugino at Brigham and Women's Hospital, Boston, Massachusetts, to observe and serve as unofficial experimental subject for noninvasive magnetic brain stimulation using MAGSTIM-200 and Cadwell MES-l0 devices (7-10 to 7-15)

1986–88   Hospital rotations as EEG technology student

1983–85   Psychiatric Resident in the State University of New York at Buffalo Affiliated Hospitals Program

1981–83   Clinical Clerkships at Niagara Falls Memorial Medical Center

1978   Exemption via Qualifying Test and Teaching Assistantship for Biomedical Psychology course during fall semester at S.G.U.S.O.M.

1978   Research Assistant in psychophysiology Lab at VA Hospital, working on project aimed at correlating EEG with psychological variables (especially depression) under auspices of University of Utah, Bioengineering Department, Salt Lake City, Utah

1976   Psychosurgeons Conference in Gottingen, West Germany (mid-July)

1974   Research Assistant in Artificial Vision for the Blind Project, Artificial Organs Program: "The Artificial Eye," University of Utah (autumn) under the auspices of Dr. William H. Dobelle, who is cited in the 2005 *Guinness Book of World Records* for "earliest successful artificial eye" (page 20, under "Medical Phenomena")

1972   Participant in Psychophysiology experiments involving EEG and feedback at the Institute of Living in Hartford, Connecticut

1971   Volunteer Remedial Reading Tutor for disadvantaged children in Hartford

1970   Discussion group leader at Red Cross Training Camp at Manilus, New York (summer)

1969–70   Remedial Reading teaching assistant during summer sessions at the Park School of Buffalo

1969   Employed as Remedial Reading tutor for elementary school student with reading difficulties

PREVIOUS PUBLICATIONS and one manuscript

1973   Unpublished initial manuscript done as "senior thesis" (actually Open Semester term paper) Open Semester Project, Spring 1973: Practical Implications of Learning and Cognitive Facilitation Theory; 119 pages. Done under the auspices of Dr. George W. Doten, Chairman of Psychology Dept. at that time at Trinity College (Hartford, Connecticut)

1982   How Learning Ability Might Be Improved by Brain Stimulation, *Speculations in Science and Technology*, 5 (No. 1): 51–53.

1986   Brain Stimulation to Treat Mental Illness and Enhance Human Learning, Creativity, Performance, Altruism and Defenses against Suffering, *Medical Hypotheses*, 21: 209–219.

1990   Riley-Day Syndrome, Brain Stimulation and the Genetic Engineering of a World Without Pain, *Medical Hypotheses*, 31: 201–207.

1992   Ultrasonic Antidepressant Therapy Might Be More Effective Than Electroconvulsive Therapy (ECT) in Treating Severe Depression, *Medical Hypotheses*, 38: 350–351.

1992   A Magnetic Choke-Saver Might Relieve Choking, *Medical Hypotheses*, 38: 349.

1993   A Proposed Method of Pleasure-inducing Biofeedback Using Ultrasound Stimulation of Brain Structures to Enhance Selected EEG States, *Speculations in Science and Technology*, 16 (No. 1): 78–79.

1995   Waiting Hopefully, *Western New York Mental Health World*, 3 (4), Winter: 14. Written under pen name Nemo T. Noone.

2006   *How Everyone Could Be Rich, Famous, Etc.*, Trafford Publishing, 2006: 240 pages.

OTHER

1.   National Certificate of Merit, 1969; New York State Regents Scholarship, 1969; Merit's Who's Who Among American High School Students, 1968–69; Advanced Placement in American History and English, 1969–70

2.   Member of the American Society of Electroneurodiagnostic Technologists (ASET), 1987

3.   Member of the American Medical Association

4.   Graphic Controls Corporation EEG Technology Award, 1988

5.   Student Representative to the EEG Technician Curriculum Advisory Committee at N.C.C.C., 1988–89

6.   Member, EEG Technician Curriculum Advisory Committee, 1989–92

7.   General Member of the American Institute of Ultrasound in Medicine (A.I.U.M.), 1993. (Was invited to join on the basis of publication #4, listed above.)

8.   Member of the American Mensa, Ltd.

9.   Member's Presentation, on Saturday, May 28, 1994, at Western New York Mensa Society's annual regional gathering, Radisson Hotel, Niagara Falls, New York; Subject: "Noninvasive Brain Pacemaker … a Real-Life Thinking Cap"

3.   Guest/Alumni Speaker, on Thursday, February 8, 1996, at the Park School of Buffalo, Snyder, New York; subject: book in progress/title of synopsis: *How Everyone Could Be Financially Stress-Free, Famous, Sexually Liberated, Effort-Free, Well Educated, Healthy, Deathless, Unselfish, Etc., via Electromagnetic, Ultrasonic or other kinds of Brain Stimulation/Pacemakers and Internet of Mind Particles, Genetic Self-Engineering, Etc.*; title of book: *How Everyone Could Be Rich, Famous, Sexually-Liberated, Etc., Etc.*

4.   Donated oil painting, 4' x 5', done during summer, autumn & winter, 1970–71, titled "To Be Whatever We Want to Be," to Trinity College, Hartford, Connecticut, in early 1971. Displayed in front hallway of Life Sciences Building for approximately ten years.

5.   Historical Committee member, St. Louis Catholic Church, 1995—

6. Member, Fathers' Rights (Organization) of W.N.Y., 1994 — (daughter born in 1983).

7. Member, Society for the Scientific Study of Sex, 1989; renewed '97.

8. Member, Inventors' Alliance of Canada/America, 1997.

9. Member's Presentation for Inventors' Alliance, "How Learning and Work Could Be Made Both Thoroughly Effortless and Intensely Enjoyable: via Noninvasive (i.e., nonsurgical, non-implanted, external, extra-cranial) Brain (Pacemaker) Stimulation," April 2, 1997.

10. Unofficial (unpaid) companion to an elderly blind man; 1993—present (2009+).

UNIVERSITY OF CALIFORNIA, IRVINE

BERKELEY • DAVIS • IRVINE
• LOS ANGELES •
RIVERSIDE • SAN DIEGO
• SAN FRANCISCO •
SANTA BARBARA •
SANTA CRUZ

UCI COLLEGE OF MEDICINE   IRVINE, CALIFORNIA
92717
DEPARTMENT OF RADIOLOGICAL SCIENCES
DIVISION OF PHYSICS AND ENGINEERING

June 26, 1992

To Whom It May Concern:

This will confirm that Dr. Lewis Mancini visited my ultrasound research laboratory the week of June 22, 1992. He viewed in some detail the ongoing experiments conducted by Mr. Christos Georgiades and supervised by Dr. Patricia Rinaldi and myself involving the use of ultrasound pulses to alter communication pathways in brain tissue. These experiments, although conducted in-vitro on sections of rat hippocampus, could potentially be applied in-vivo to human subjects. In humans, such experiments could lead to a better understanding of brain function and might someday aid in the diagnosis and treatment of a variety of brain disorders.

During the week of June 22nd, we had extensive discussions with Dr. Mancini concerning the use of ultrasound to affect the way neurons in the brain respond. Our discussions ranged from purely scientific to the very practical. I found Dr. Mancini to be a bright and very engaging person. He is clearly interested in our ultrasound research from both a scientific and personal point of view. I would welcome regular visits by him to our lab and continued discussions with him on a long-term basis; I believe such discussions would be stimulating and productive and could well lead to new insights and discoveries of both a scientific, as well as a practical nature.

   (Signed)

Joie Pierce Jones
Professor of Radiological Sciences

JPJ:pc

September 13, 1989

To whom it may concern:

This is to certify that I am cooperating with Dr. Lewis Mancini in a graduate level independent study and research project concerning the design, development and use of a noninvasive stimulator of focal regions of cns and peripheral nerve tissue. The device as proposed by Dr. Mancini will include highly focal magnetic and ultrasonic field components.

I have reviewed Dr. Mancini's background in biomedical engineering and medical sciences and find him thoroughly qualified to participate as a principal in this study at the level commensurate with graduate credit in my discipline (Electrical Engineering).

Sincerely,

(Signed)

Peter D. Scott
Associate Professor

Graphic Controls

Division

September 26, 1988

Mr. Lewis S. Mancini

Dear Lewis:

Congratulations on being the recipient of the 1988 Graphic Controls Corporation Electroencephalography Award for Academic and Technical Excellence.

A commemorative plaque is enclosed to acknowledge this achievement. A One Hundred Dollar U.S. Savings Bond has been sent to you.

We at Graphic Controls are pleased to provide this award for your distinguished achievement.

Very truly yours,

(Signed)

Al Barber
Product Marketing Manager
Patient Data Supplies

AB/dc:3925u

PennState

ENGINEERING SCIENCE & MECHANICS SEMINAR

presents

SPEAKER:

LEWIS S. MANCINI
State Univ. of New York at
Buffalo School of Medicine
Biophysics/EEC Program

TOPIC:
*or a Real-Life
Thinking Cap"*

"Noninvasive Brain Pacemaker

DATE:

Wednesday, February 17, 1993

TIME:

3:35 P.M.

LOCATION:

Room 314 Hammond Bldg.

REFRESHMENTS WILL BE SERVED IN ROOM 226 HAMMOND BEGINNING AT 3:00 P.M.

PENNSTATE

Department of Engineering Science
and Mechanics

The Pennsylvania State University

Akhlesh Lakhtakia
Associate Professor

February 23, 1993

Dr. Lewis Mancini

Dear Lewis:

It was my sincere pleasure to have met you last week and listened to your ideas on the ultrasonic stimulation of the human brain. This is a potentially exciting idea that needs more work on. If it is successful, I am quite sure it will lessen performance anxieties and thereby lessen conflicts between people.

Please keep me occasionally informed about your work. With my best wishes, I remain

Sincerely yours,

(Signed)

Akhlesh Lakhtakia
Fellow, Optical Society of America
Fellow, Leonhard Center for Innovation
Editor-in-Chief, Speculations in Science and Technology

An Equal Opportunity University

niagara falls memorial medical center

December 13, 1982

C. Timothy Golumbeck, M.D.
Director of Residency Training
Department of Psychiatry
Erie County Medical Center

    Re: Lewis Mancini

Dear Doctor Golumbeck:

I have known Dr. Mancini since early October, 1981, when he came for a six-week psychiatric rotation at our inpatient unit of Niagara Falls Community Mental Health Center. He came back for another rotation in late October, 1982, for a period of four weeks. He apparently has been interested in the field of Psychiatry for a long period of time and he states that he was involved in, or ran, Research concerning this field.

He impresses me as a very conscientious young man, having a high moral standard, and very energetic and enthusiastic about certain areas which he is interested in. He is also immaculate and punctualistic. He also willingly admits some of his weaknesses, which may have influenced him to be more interested in the research area than direct patient care.

I am particularly impressed by his openness and some fascinating ideas he can come up with. With further training and support, I'm sure that he will be able to contribute to the better understanding and further development of Psychiatry in the future, particularly in the field of Psychiatric Research.

I wish him to be accepted into the program for which he wants to be trained.

Sincerely,

NIAGARA FALLS COMMUNITY
  MENTAL HEALTH CENTER

(Signed)

J. S. Rhee, M.D.
Psychiatrist

JSR:ve

DIVISION OF LIFE SCIENCES

TO WHOM IT MAY CONCERN

Lewis Mancini, MD, has recently completed an AMA accredited Electroencephalography Program that is jointly sponsored by Niagara County Community College and the State University of New York at Buffalo, School of Medicine. As a student in the EEG Program, he was academically the highest in the class. He is the recipient of the Graphics Control Award in Neurodiagnostic Technology (1988) for Clinical and Academic Excellence. Dr. Mancini holds a Bachelor of Science degree from Trinity College (Hartford, Connecticut) and an MD degree from St. George's University School of Medicine (Grenada, West Indies).

His interests appear to lie in the areas of bioengineering, neurology, and psychiatry. Lewis completed a special independent project on Riley-Day Syndrome, resulting in a paper titled "Riley-Day Syndrome, Brain Stimulation, and the Genetic Engineering of a World Without Pain of Any Kind." Some thought-provoking ideas emerged from this study. He has recently published an article in <u>Medical Hypotheses</u> titled "Brain Stimulation to Treat Mental Illness and Enhance Human Learning, Creativity, Performance, Altruism, and Defenses Against Suffering."

Dr. Mancini always dressed in a neat and conservative fashion, always came to class on time, was well prepared, and contributed actively to the discussions. In the context of his rotations at the four hospitals in the EEG Program, he had a uniformly positive rapport with both clinical personnel and patients.

I feel he has both the academic qualifications and professional maturity to engage in independent research and I whole-heartedly support this endeavor.

If I may be of further assistance relative to Dr. Mancini's credentials, please do not hesitate to call or write.

Sincerely,

(Signed)

Nicholas J.T. LoCascio, Ph.D.
Assistant to the Vice President of
    Academic Affairs – NCCC
Clinical Instructor, Neurology
    State University of NY at Buffalo
    School of Medicine

Indianapolis Center for
Advanced Research, Inc.
May 27, 1987

Lewis S. Mancini, M.D.

Dear Dr. Mancini:

Your survey of articles pertinent to the possibility of focal excitation of CNS seems to have been both perceptive and accurate. Since what you are proposing to accomplish is a much sought after and worthwhile goal it is appropriate to applaud your persistence and analysis. No noninvasive method to my knowledge has yet been shown to accomplish what you seek to do.

My first comment is to state that I think it is only a matter of time before the goal is achieved. It may be achievable with careful experimental application of known technology which you have reviewed in your letter. The size of the stimulated volume with present technology if it works would undoubtedly be limited to several cubic millimeters.

One would of course not start out to prove a given methodology if it involved ultrasound by leaving the skull in place in experimental animals. Once the method was proven the skull complication could then be addressed. This is my answer to your first question.

Since the second question requires a high degree of speculation I prefer to answer question one in as firm and positive manner as I can and hope that you can take definitive steps to carefully apply basically known technology to demonstrate feasibility. I would place the probability of a reasonable measure of success in the hands of careful experimenters as well over 50%.

Sincerely,

(Signed)

Francis J. Fry
Senior Scientist

FF:kk

Medical Center   Buffalo, NY

**Veterans
Administration**

In Reply Refer To:

April 15, 1986

TO WHOM IT MAY CONCERN:

    RE: Lewis Mancini, M.D.

I'm writing this letter on behalf of Dr. Lewis Mancini. I had the opportunity to work with Dr. Mancini during his rotation in Consultation/Liaison Psychiatry Service at the Buffalo VA Medical Center from January to June of 1985.

Dr. Mancini demonstrated initiative, cooperative team spirit and empathy in dealing with patients, families and staff. He was highly motivated in pursuing his own research.

I do hope you will consider his application seriously.

Sincerely,

(Signed)

Fern E. Beavers, R.N., M.S.
Clinical Nurse Specialist/Psychiatry

"America is #1 — Thanks to our Veterans"

UNIVERSITY AT BUFFALO

STATE UNIVERSITY OF NEW YORK

Department of Psychiatry

School of Medicine
Faculty of Health Sciences
VA Medical Center
3495 Bailey Avenue
Buffalo, New York 14215
(716) 862-3670

November 25, 1985

TO WHOM IT MAY CONCERN:

Dr. Lewis Mancini rotated in Consultation/Liaison Psychiatry Service of the Buffalo VAMC under my supervision from January to June of 1985 as part of his II year Psychiatry Residency Training. I found him to have adequate psychiatric knowledge for his level. He was clinically responsible and followed up his patients. He showed initiative and creativity, leaning more toward research of the use of electromagnetic forces to influence thinking processes and intelligence which he said he plans to pursue further. He showed adequate learning ability and was eventually able to perform his functions autonomously and confidently. He presented to me as being quite motivated to pursue the goals he set up for himself in life.

(Signed)

A. E. PENETRANTE, M.D.
Staff Psychiatrist
C/L Service
Buffalo VAMC